Karl Zilles

The Cortex of the Rat

A Stereotaxic Atlas

With 130 Figures

Springer-Verlag
Berlin Heidelberg NewYork Tokyo

Prof. Dr. KARL ZILLES
Anatomisches Institut der Universität Köln
Joseph-Stelzmann-Straße 9
D-5000 Köln 41

Cover photo: see page 84, Figure 62a

ISBN-13: 978-3-642-70575-5 e-ISBN-13: 978-3-642-70573-1
DOI: 10.1007/978-3-642-70573-1

Library of Congress Cataloging-in-Publication Data
Zilles, Karl J., 1944. The cortex of the rat.
Bibliography: p. 113. Includes index.
1. Cerebral cortex – Anatomy – Atlases. 2. Rats – Anatomy – Atlases. I. Title.
QL938.C46Z55 1985 599.32′33 85-14702

Softcover reprint of the hardcover 1st edition 1985

2125/3130-543210

Contents

Introduction

The rat brain is the most widely used animal model in neurobiology. The frequency of its use underlines the need for stereotaxic atlases of the rat brain. The first attempt to fulfill this requirement was made by Krieg (1946c), but his atlas contains only very rough data based on 30 schematic drawings of coronal sections. The drawings are exaggeratedly schematic and significantly distorted. This, together with the fact that Krieg delineated far too few structures in the di- and rhombencephalon, makes the accurate placement of electrodes difficult. Accurate electrode placement is essential for advanced techniques in modern [1] neuroanatomical, neurophysiological, and neurochemical studies. Nevertheless, Krieg's work represents the first delineation of cortical areas in the rat. These delineations were more elaborately produced and partly corroborated by experimental work by Krieg (1946a, b, 1947). Despite its shortcomings, this fundamentally valuable series of reports greatly influenced experimental approaches in the four decades following its publication. The work of König and Klippel (1963) was another landmark in the development of stereotaxic atlases of the rat brain. This atlas is one of the most widely used works of reference on stereotaxic coordinates of the rat brain. Some difficulties, however, arise from the drastic changes there have been in the accepted parcellation of the thalamus, septum, and amygdala since the publication of this atlas. In addition, stereotaxic coordinates for structures of the pons, medulla oblongata, and cerebellum are lacking. A further deficiency is the absence of cortical areal boundaries. Yet another source of considerable electrode placement error is that the work was based on the rather idiosyncratic use of young (150 g) female rats. Older and heavier rats are used for most current research (cf. Paxinos and Watson 1982). Some additional subcortical structures are delineated in the atlas by Pellegrino and Cushman (1979). Stereotaxic coordinates for newborn and growing rats have been established by

Sherwood and Timiras (1970) and by Heller et al. (1979). These works also present data on subcortical structures only. Finally, the architectonic work of Wünscher et al. (1965) deals exclusively with the pons and medulla of the rat.

Paxinos and Watson (1982) made a significant advance in that they used the flat-skull position, which can be reproduced with adequate precision. Bregma, lambda, and the midpoint of the interaural line can be used as reference points and are described in the figures. The coordinates are calculated for male Wistar rats weighing 250–350 g, that is to say, for the weight and strain most widely used in experimental work. Fresh brains frozen in the skull were cut and sections presented at 0.5-mm intervals in the coronal, sagittal, and horizontal planes. With this technique the usual distortions caused by histological shrinkage have been avoided. Cytoarchitectural, hodological, and histochemical data were the basis for the delineation of subcortical structures. The presently accepted organizational schemes were maintained throughout these delineations. Craniometric and stereotaxic data for correction of numerous distances, as in the case of other rat strains or younger animals, are also given. This atlas by Paxinos and Watson (1982) is the most comprehensive and advanced work of reference for stereo-
[2] taxic studies on subcortical structures of the rat brain.

The primary reason for the present atlas, therefore, is not to present even more data on stereotaxic coordinates of subcortical structures, but to give stereotaxic data on *cortical* areas of the rat brain. This aspect is not considered by Paxinos and Watson (1982) or in any of the earlier atlases. It seems that cortical delineations are in great demand because advanced techniques in neuroanatomy, neurophysiology, and neurochemistry of the cortex provide data whose interpretation could be ambiguous without precise knowledge of the cortical areas involved. Very often it is totally inadequate to make a crude subdivision of cortical tissue into neo- and allocortex or even into frontal and occipital cortex. Numerous architectural, electrophysiological, histochemical, metabolic, psychological, and other studies have unequivocally established the heterogeneity of the different cortical areas. These results should be accessible for topographic analysis by scientists who are inexperienced in the difficult and often controversial problems of cortical parcellation. In some cases, the experimental methods used or staffing and instrumentation restrictions of the particular laboratory may make the precise identification of cortical areas impossible. This atlas presents stereotaxic coordinates for cortical areas of the rat brain in close analogy to the Paxinos and Watson (1982) atlas, which is gaining acceptance as the standard work of reference

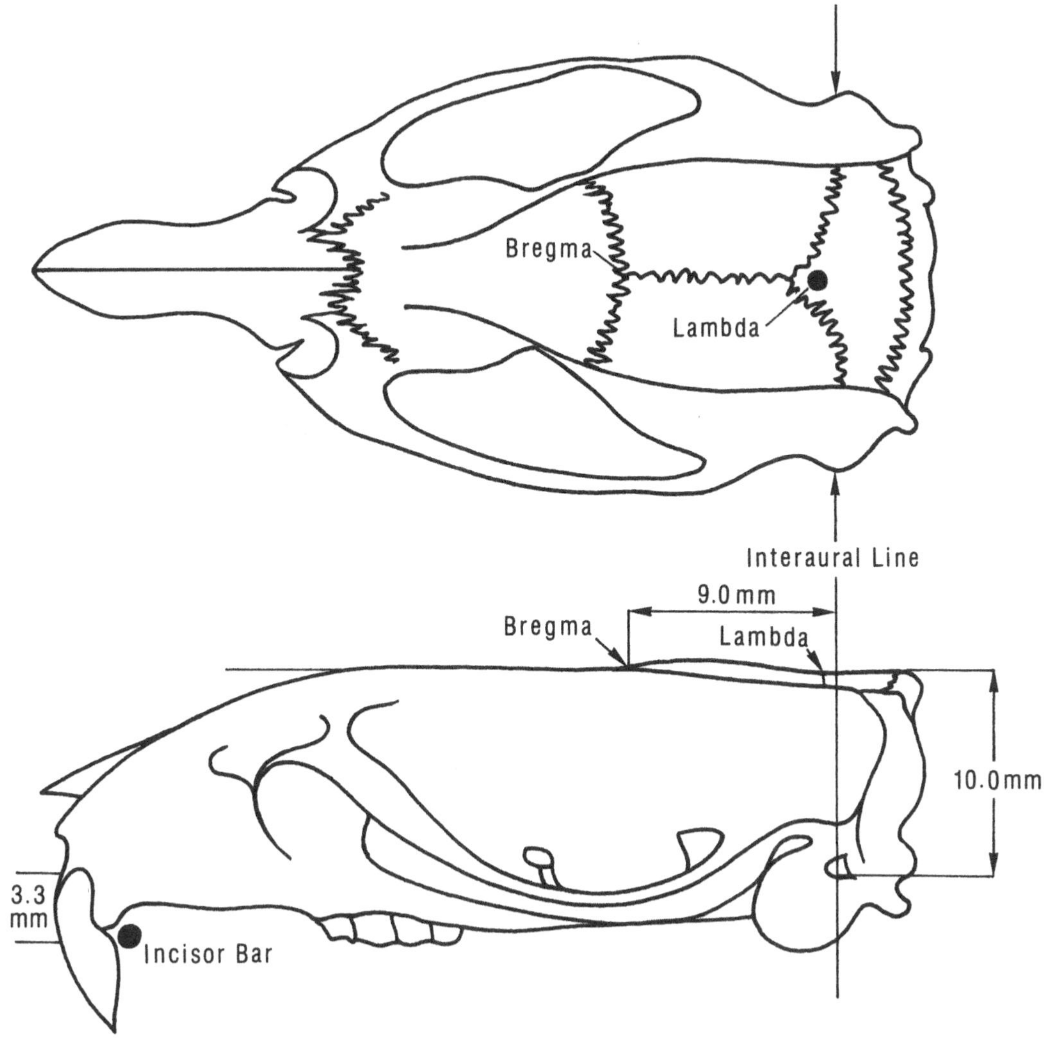

Fig. 1. Dorsal and lateral views of a rat skull with the positions of bregma, lambda, and the interaural line. Lambda is about 0.3 mm anterior to the coronal plane passing through the interaural line. (After Paxinos and Watson 1982)

on stereotaxis. The original histological sections of this atlas have been used in the present work for the delineation of cortical areas. Subcortical structures have been delineated in the very few instances in which they are helpful or necessary for orientation. This arrangement has several advantages. First, it makes republication of low-power micrographs of the original histological sections on which the plates are based unnecessary, because these can be found in the Paxinos and Watson (1982) atlas. This allows space for the reproduction of low- and medium-power photomicrographs of histological sections specifically prepared for the demonstration of cortical structures of the rat and their variability in structure and position. Second, a fairly complete description of the subcortical and cortical structures of a representative brain can be established, because all the information on subcortical structures found in the Paxinos and Watson (1982) atlas can be transferred directly to this atlas of cortical structures if necessary.

The main features of this atlas are:

1. Line drawings of 30 coronal (distance: 0.5 mm), 12 horizontal (distance: 0.5 mm), and 2 sagittal sections with stereotaxic coordinates based on the flat-skull position. The boundaries of 44 isocortical and allocortical areas are delineated and cortical maps are graphically reconstructed by orthogonal projection.
2. The frontal and occipital poles are shown from varying angles of inspection. These figures were based on computer-aided reconstruction and rotation of the line drawings.
3. Structures are delineated on the basis of Nissl-, myelin-, and acetylcholinesterase-stained sections from 12 additional rat brains.
4. The delineation of cortical structures is based on a quantitative approach using television image analyzers.

[4] 5. Low- and medium-power photomicrographs illustrate the overall appearance and laminar structure of the different cortical areas.

6. Cortical maps of different rat brains demonstrate the intraspecific variability in regional patterns.

Nomenclature

A review of the nomenclature for cortical areas in the rat reveals a confusing variety of terms and terminological systems. Brodmann's (1909) nomenclature is designed mainly for primates. Although Brodmann personally used this numerical system to describe many nonprimate species, including rodents, the system seems to be inadequate for such use, because problems of comparative anatomy and functional equivalence between primates and rodents are unresolved. One shortcoming is evident insofar as Brodmann (1909) incorrectly identified the auditory cortex in rodents. He presented a peculiar areal pattern in the temporal region. Krieg (1946, 1947) revised and expanded Brodmann's nomenclature for use with rats. A second, more biologically relevant problem arises from the implied comparability between the cortical areas of primates and rodents when the same areal terms are used to describe both. In most cases this problem of homology of cortical areas is unresolved. Many other systems of nomenclature have been designed for special studies, but these are not generally applicable outside the specialized fields for which they were designed.

A purely functional nomenclature is not appropriate, because clearly analyzed functions have been found for only some of the anatomically defined areas. A nomenclature system which is neutral, and therefore open to future supplementation without alteration to the entire system, has been used in an earlier review (Zilles and Wree 1985) of the areal and laminar pattern of the rat cortex. This alphanumeric system uses abbreviations to denote topographic regions (Fr = frontal cortex; Par = parietal cortex; Oc = occipital cortex; Cg = cingulate cortex, etc.) or to denote traditionally used terms (CA = Cornu ammonis; RSA = retrosplenial agranular cortex). Numbers are used to supplement these abbreviations wherever necessary. In the isocortex, for example, primary and supplementary somatosensory cortices are denoted as Par1 and Par2, respectively. The secondary visual cortex requires further parcellation. This is indicated by additional alpha indices (Oc2M = medial secondary visual cortex; Oc2MM = medial part of Oc2M; Oc2ML = lateral part of Oc2M). The cortical maps in the present volume correspond to maps previously published in papers by Zilles et al. (1984) and Zilles and Wree (1985). These papers present detailed information on delineation and identification problems. The area described as 29D by Zilles et al. (1984) is defined in this atlas as the medial part of the secondary visual cortex and is therefore described as Oc2MM. Axonal transport studies with HRP (personal unpublished studies) suggest this revised identification.

Coronal, Horizontal, and Sagittal Sections in Stereotaxic Coordinates

The line drawings (Figs. 2–45) which include stereotaxic coordinates, are based on the original histological sections of adult male Wistar rats published by Paxinos and Watson (1982). The maps in Figs. 46–49 are also prepared from these drawings by orthogonal projection.

The brains of these rats were either frozen within the skulls (sagittal and horizontal series) or removed from the calvarium and then frozen (coronal series). Brains were frozen with dry ice or CO_2 and were sectioned at 40-µm intervals. They were then stained with either cresyl violet or AChE (Koelle and Friedenwald 1949; Lewis 1961). A detailed description of these procedures is given in the Paxinos and Watson (1982) atlas. The boundaries of cortical areas in the present study were determined in the original Nissl- and AChE-stained sections with a quantitative morphological method using a computer-controlled image analyzer (Zilles et al. 1980, 1982, 1984; Zilles and Wree 1985). The equipment used to carry out the measurements on Nissl sections consists of a Micro-Videomat 2 image analyzer (Zeiss, Oberkochen, West Germany) linked to a Wang 2200 MVP computer. A gray level index (GLI) was determined by an automatic scanning procedure. In the present investigation the GLI is the amount of the areal proportion of Nissl-positive particles in a measuring field of defined size (30 × 30 µm or 50 × 50 µm). A previous study has shown that the GLI correlates with the nerve cell-packing density (Wree et al. 1982), even though glial cell nuclei are included by this measurement. Since the glial cell-packing density makes a constant contribution to the GLI values (Wree et al. 1982), it does not affect the GLI-to-neuron packing density ratio. The results of these measurements are presented in computer plots, in which the regions of lower or higher GLI values are plotted with lower or higher densities (Figs. 55–67). The GLI values ranging from 0 to 100% are classified in five different ranges.

Additional information on cellular morphology which cannot be shown by the GLI measurement has been included. This was based on a detailed inspection of the original histological sections. This qualitative analysis of morphological features, combined with the quantitative analysis using the GLI, provides a complex aspect of laminar pattern based on a reliable quantitative method. Both analyses served as the basis for the delineation of areal and laminar borders.

The investigation of Nissl-stained sections was further supplemented, in many cases, by measurements of the gray value distribution in alternate AChE-stained sections. These gray value measurements were generated by an IBAS 2 (Kontron, Munich, West Germany) image analyzer. In these analyses, the quantitatively described pattern of AChE activity distribution was used as an additional criterion for the delineation of cortical areas and layers.

Detailed information concerning stereotaxic procedure, use of the flat-skull position, and variability of craniometric data is provided in the atlas by Paxinos and Watson (1982). Figure 1 summarizes the most important landmarks for the stereotaxic coordinates. In summary, the procedure is:

Anesthesia is initiated with halothane vapor, followed by [8] an IP injection of barbiturate. For adult rats (350 g body weight) a solution consisting of 1 ml Nembutal (equivalent to 60 mg Na phenobarbitate) in 3 ml sterile Ringer's solution is used, 0.25 ml of this solution being injected per 100 g body weight (37.5 mg Na phenobarbitate/kg body weight). If necessary, a further 0.1 ml/100 g body weight can be injected 10 min after the initial injection. This provides deep anesthesia for 60–70 min.

The flat-skull position is reached by lowering the incisor bar of the stereotaxic apparatus to an average of 3.3 mm below horizontal zero. Horizontal zero is defined by the interaural line, which extends between the tips of the ear bars. The lambda point is located approximately 0.3 mm anterior to the interaural line. This is just caudal to the actual intersection of the lambdoid and sagittal sutures. The bregma point is defined by the intersection of the sagittal and coronal sutures.

In the *coronal drawings* (Figs. 2–31) the anteroposterior distance in millimeters from the vertical plane passing through the interaural line is displayed at the bottom left of each drawing and the anteroposterior distance from the bregma at the bottom right. The dorsoventral distance in millimeters from the horizontal plane passing through the interaural line is shown by the numbers in the left margin, and the dorsoventral distance from the horizontal plane passing through bregma and lambda on the surface of the skull is shown by the numbers in the right margin. The mediolateral distance from the interhemispheric cleft and the mediosagittal plane of the brain is given by the numbers in the top and bottom margins.

In the *horizontal drawings* (Figs. 32–43) the dorsoventral distance in millimeters from the horizontal plane passing through the interaural line is displayed at the bottom left and the dorsoventral distance from the horizontal plane passing through bregma and lambda on the surface of the skull, at top right. The anteroposterior distance in millimeters from the coronal plane passing through the interaural line is given in the bottom margin, and the anteroposterior distance from the coronal plane passing through bregma is shown in the top margin. Positions ventral to the interaural horizontal plane (position = 0) or posterior to either the interaural line or bregma are indicated by minus signs. The mediolateral distance in millimeters is shown in the right and left margins.

The *sagittal drawings* (Figs. 44 and 45) show the distance in millimeters from the midline at bottom left. The distance from the horizontal plane passing through the interaural line and the distance from the horizontal plane passing through bregma and lambda on the surface of the skull are given in the left and right margins, respectively. The distance in millimeters from the coronal plane passing through bregma is given in the top margin. The numbers in the bottom margin refer to the distance from the coronal plane passing through the interaural line.

[10]

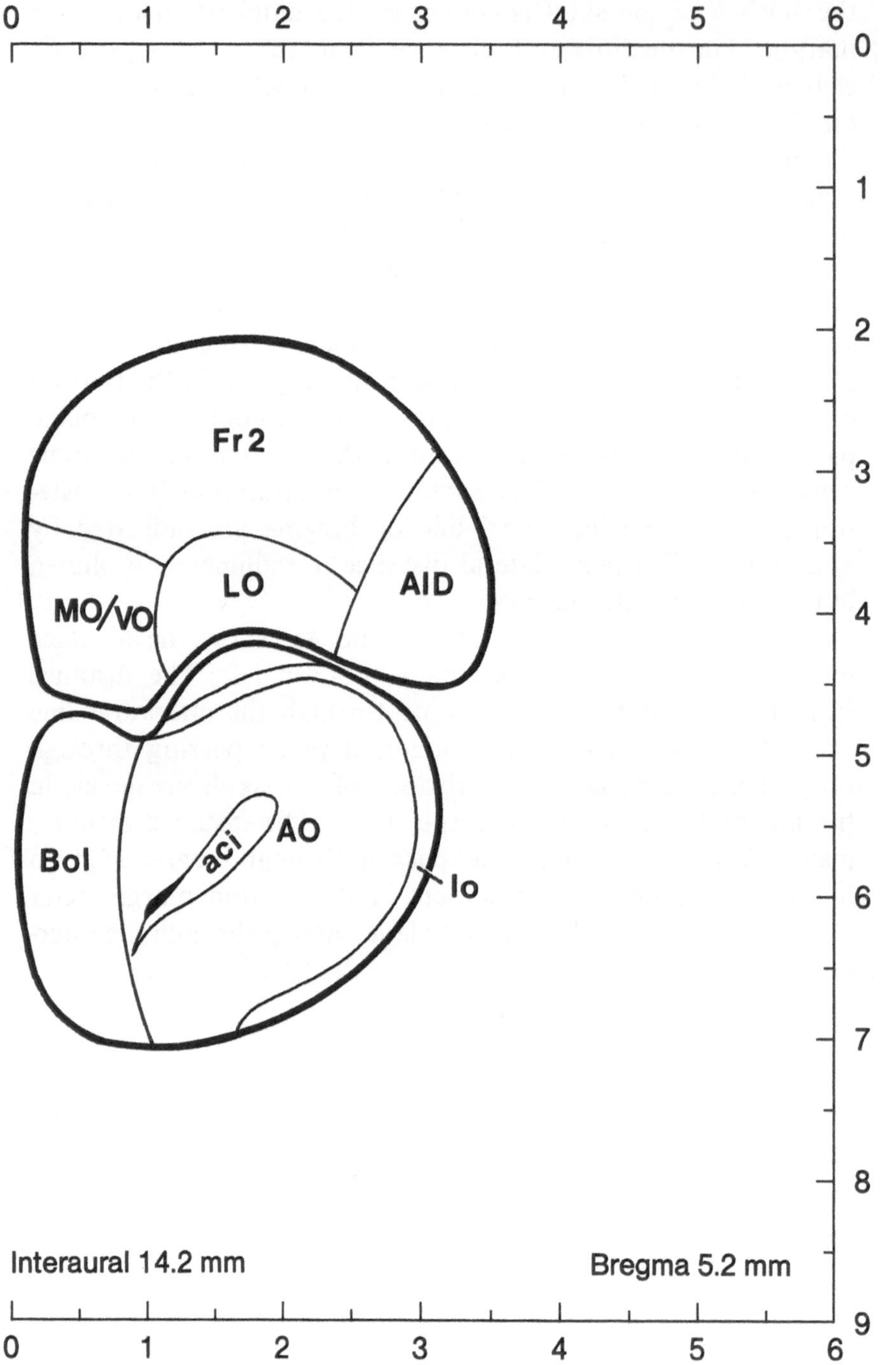
0
1
2
3
4
5
6
0
1
2
3
4
5
6
7
8
9
Fr2
MO/VO
LO
AID
Bol
aci
AO
lo
Interaural 14.2 mm
Bregma 5.2 mm
0
1
2
3
4
5
6

Fig. 2

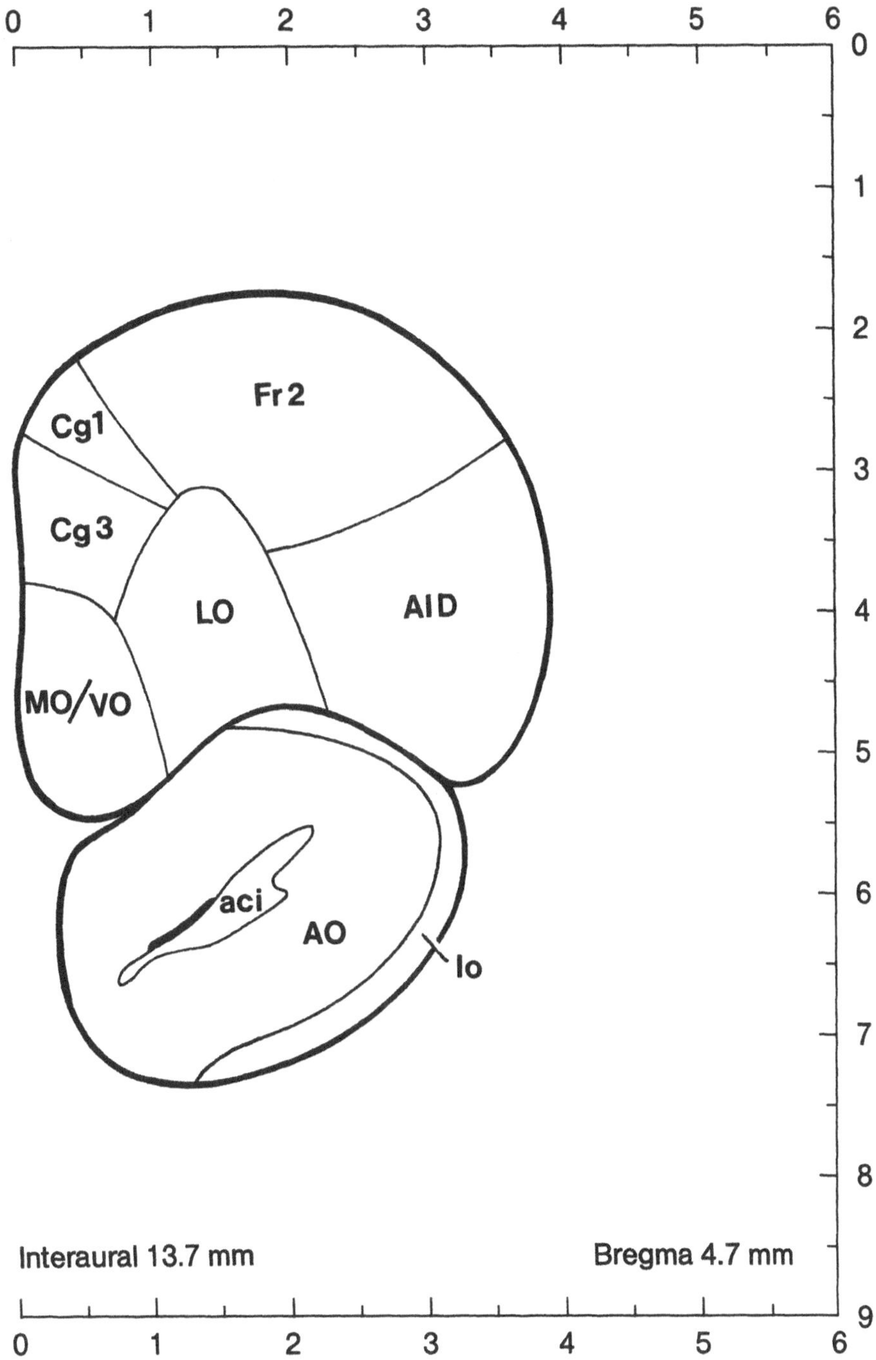
0
1
2
3
4
5
6
0
1
2
3
4
5
6
7
8
9
Fr2
Cg1
Cg3
LO
AID
MO/VO
aci
AO
lo
Interaural 13.7 mm
Bregma 4.7 mm
0
1
2
3
4
5
6

[11]

Fig. 3

[12]

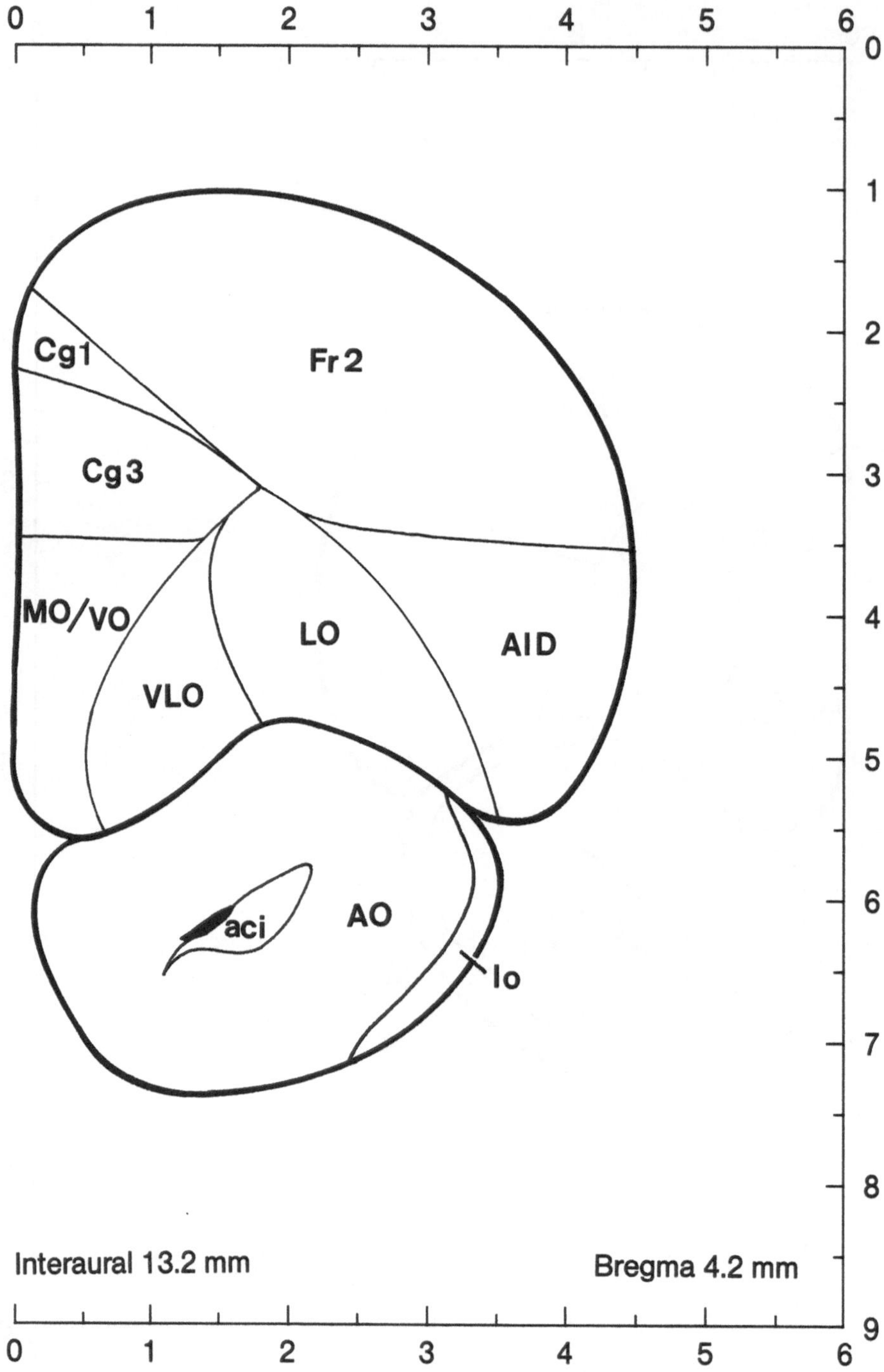

0
1
2
3
4
5
6
0
1
2
3
4
5
6
7
8
9
Cg1
Fr 2
Cg 3
MO/VO
LO
AID
VLO
aci
AO
lo
Interaural 13.2 mm
Bregma 4.2 mm

Fig. 4

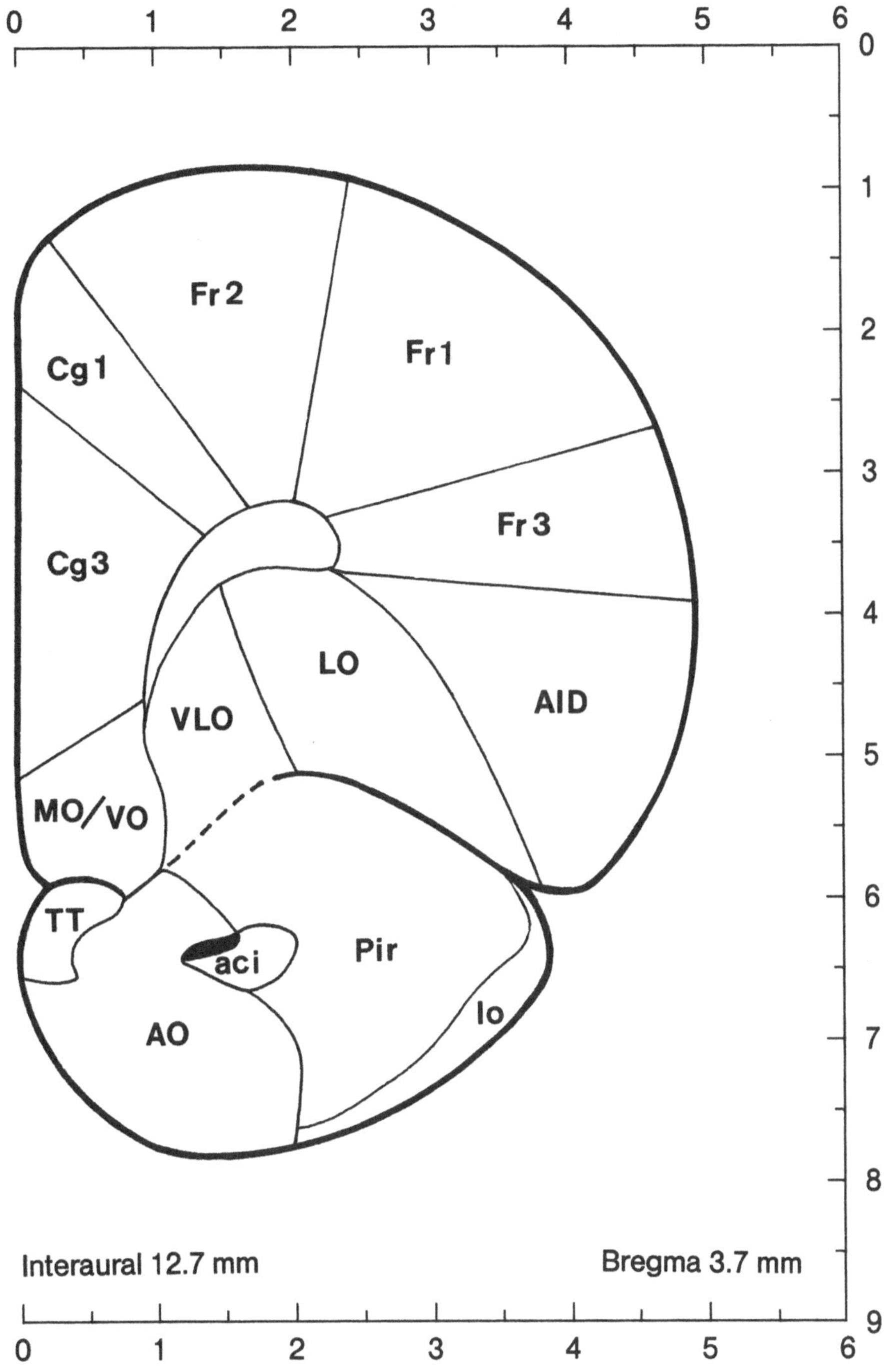
0
1
2
3
4
5
6
Fr 2
Fr 1
Cg 1
Cg 3
Fr 3
LO
VLO
AID
MO/VO
TT
aci
Pir
lo
AO
Interaural 12.7 mm
Bregma 3.7 mm
7
8
9

[13]

Fig. 5

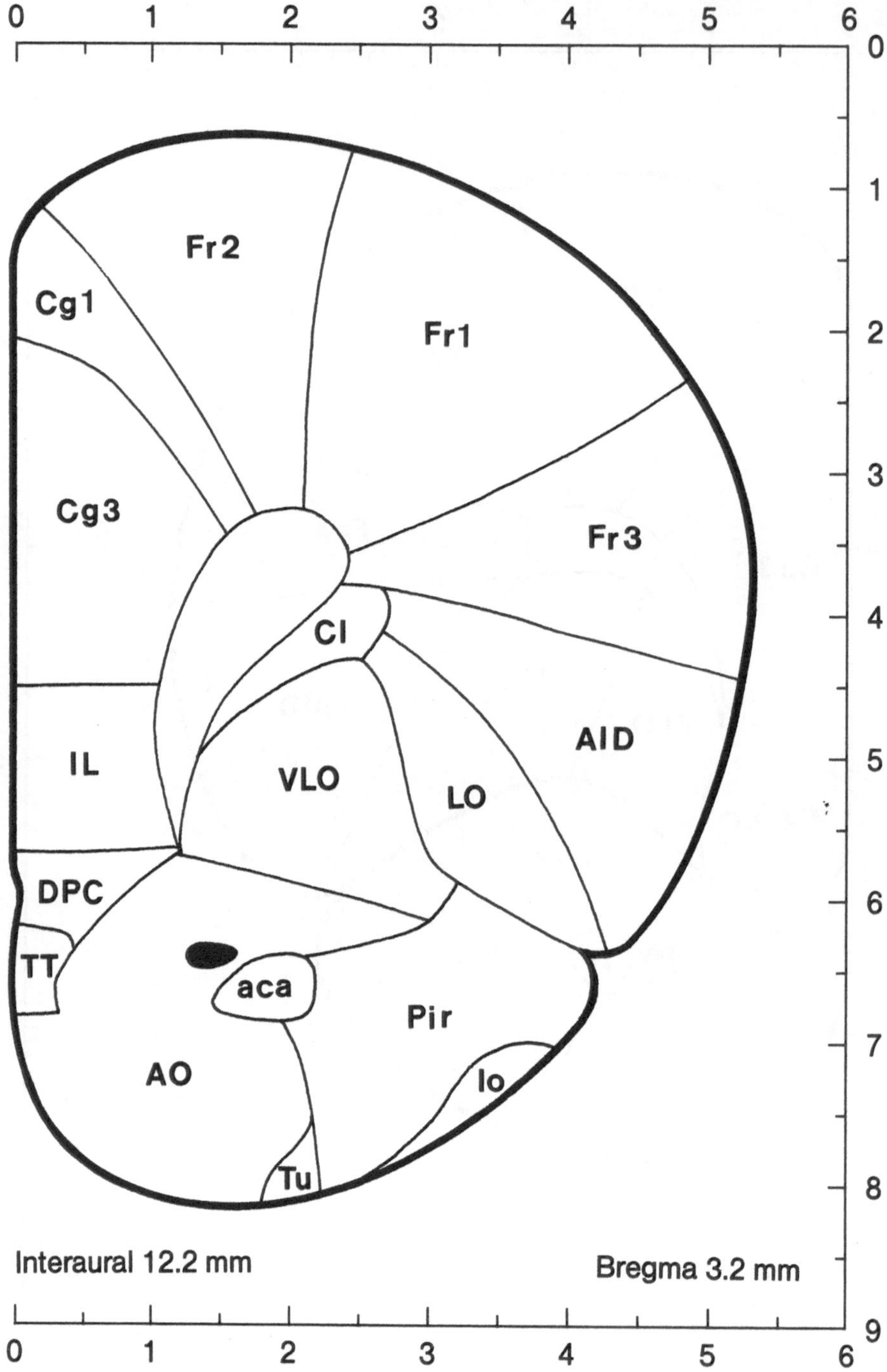

0 1 2 3 4 5 6
0 1 2 3 4 5 6 7 8 9
Fr2
Cg1
Fr1
Cg3
Fr3
Cl
AID
IL
VLO
LO
DPC
TT
aca
Pir
AO
lo
Tu
Interaural 12.2 mm
Bregma 3.2 mm

Fig. 6

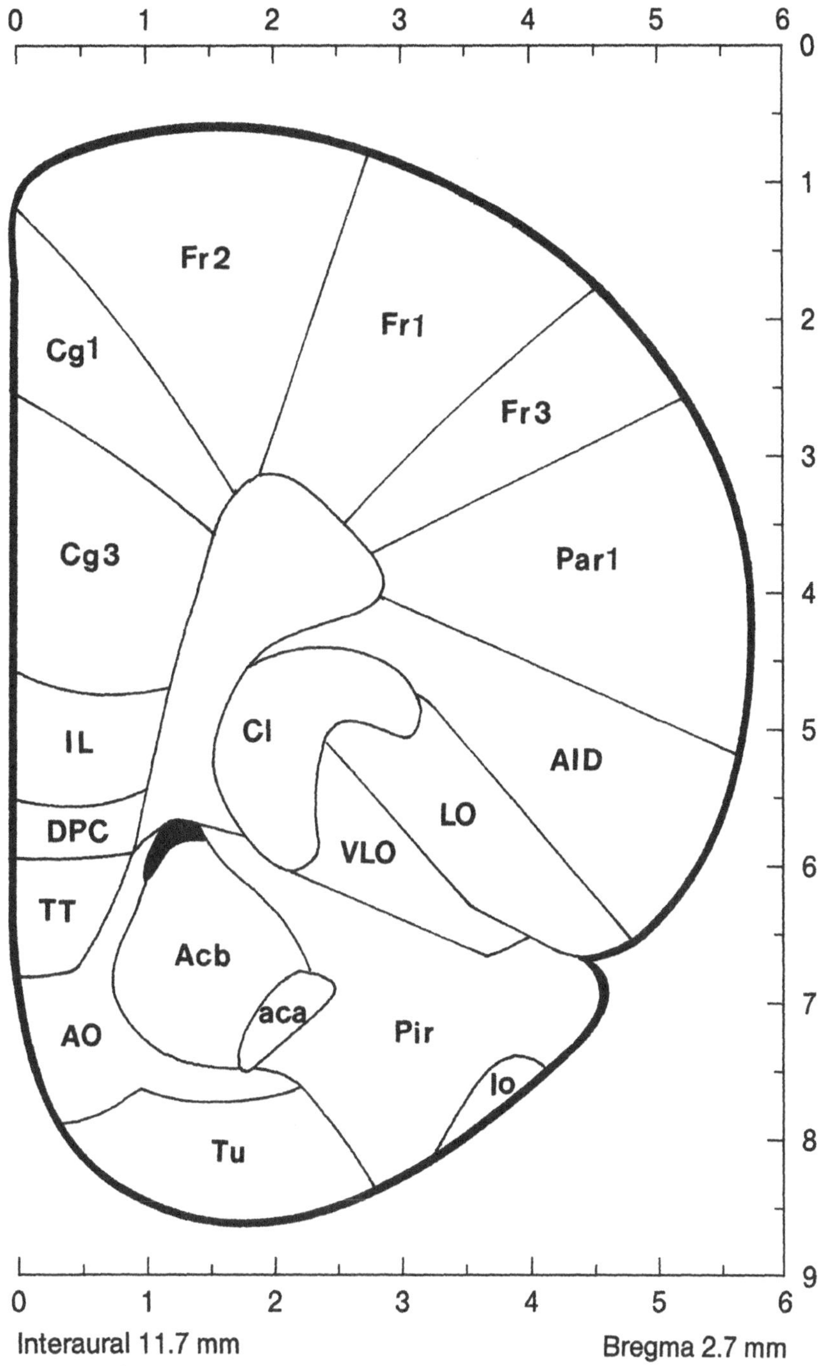
0
1
2
3
4
5
6
7
8
9
Fr2
Fr1
Fr3
Cg1
Cg3
Par1
IL
Cl
AID
LO
DPC
VLO
TT
Acb
aca
AO
Pir
lo
Tu
Interaural 11.7 mm
Bregma 2.7 mm

[15]

Fig. 7

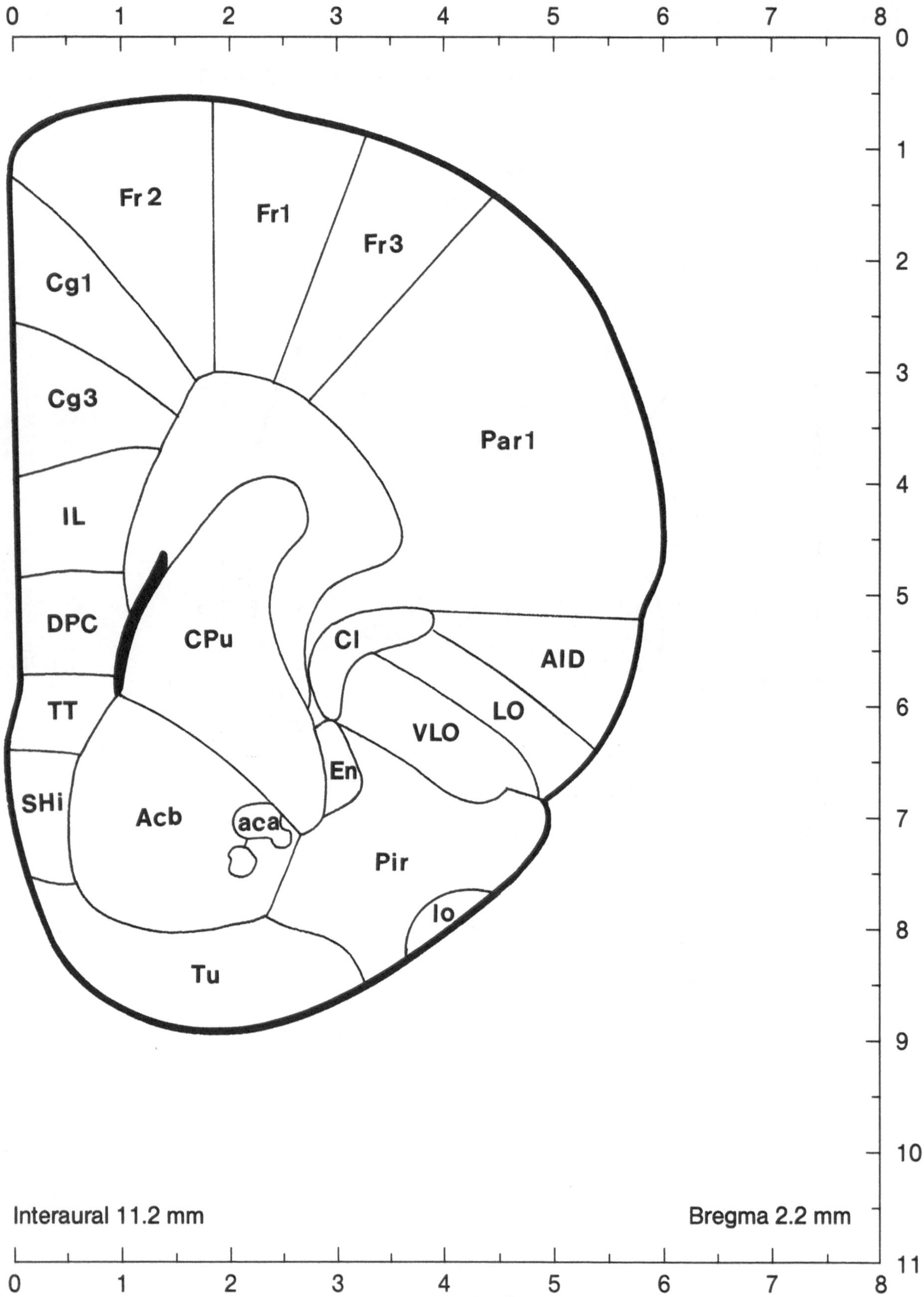
0
1
2
3
4
5
6
7
8
9
10
11
Fr 2
Fr1
Fr3
Cg1
Cg3
Par1
IL
DPC
CPu
Cl
AID
TT
LO
VLO
En
SHi
Acb
aca
Pir
lo
Tu
Interaural 11.2 mm
Bregma 2.2 mm

Fig. 8

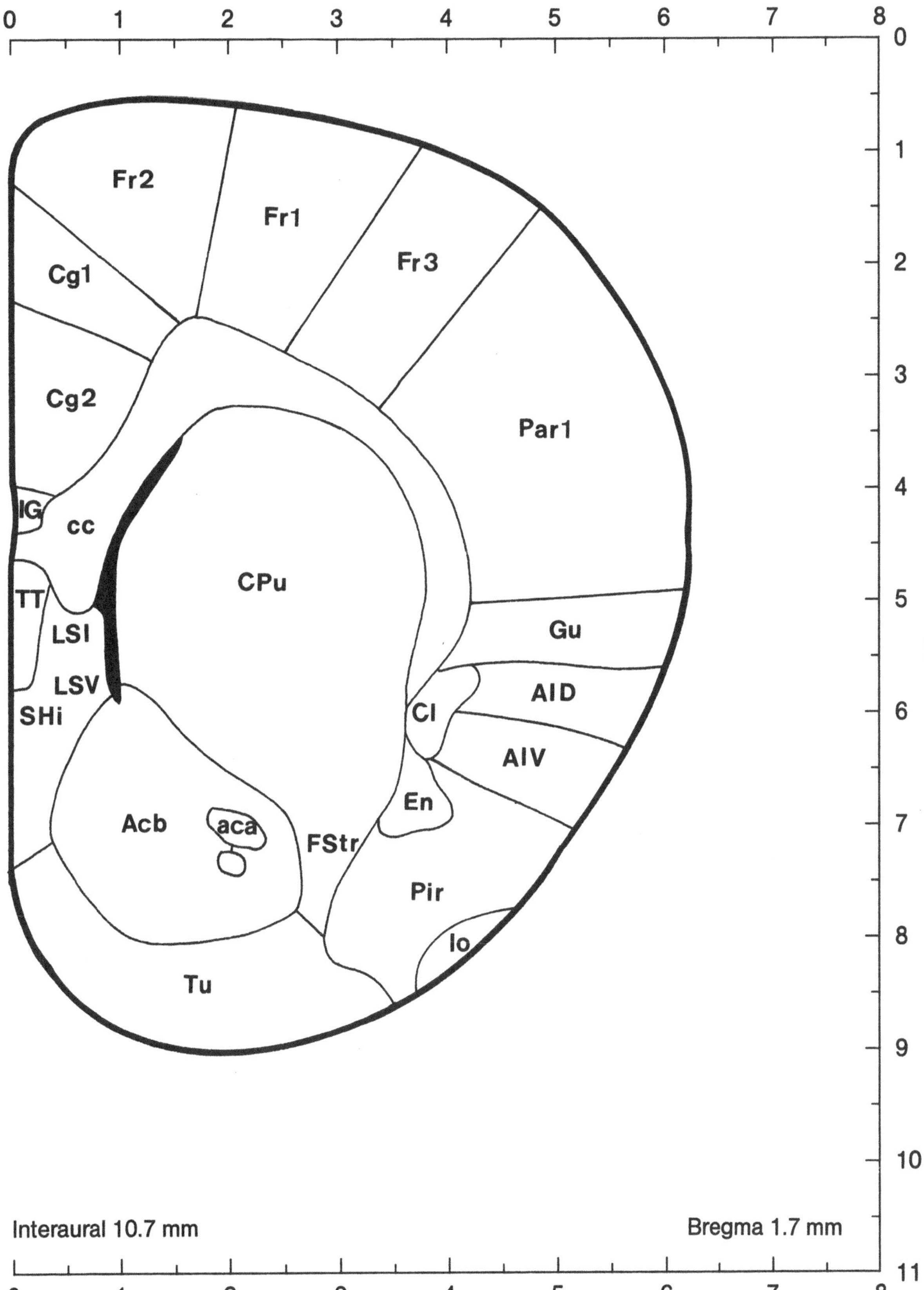
0
1
2
3
4
5
6
7
8
9
10
11
Fr2
Fr1
Fr3
Cg1
Cg2
Par1
IG
cc
CPu
TT
LSI
LSV
SHi
Gu
AID
Cl
AIV
En
Acb
aca
FStr
Pir
lo
Tu
Interaural 10.7 mm
Bregma 1.7 mm

[17]

Fig. 9

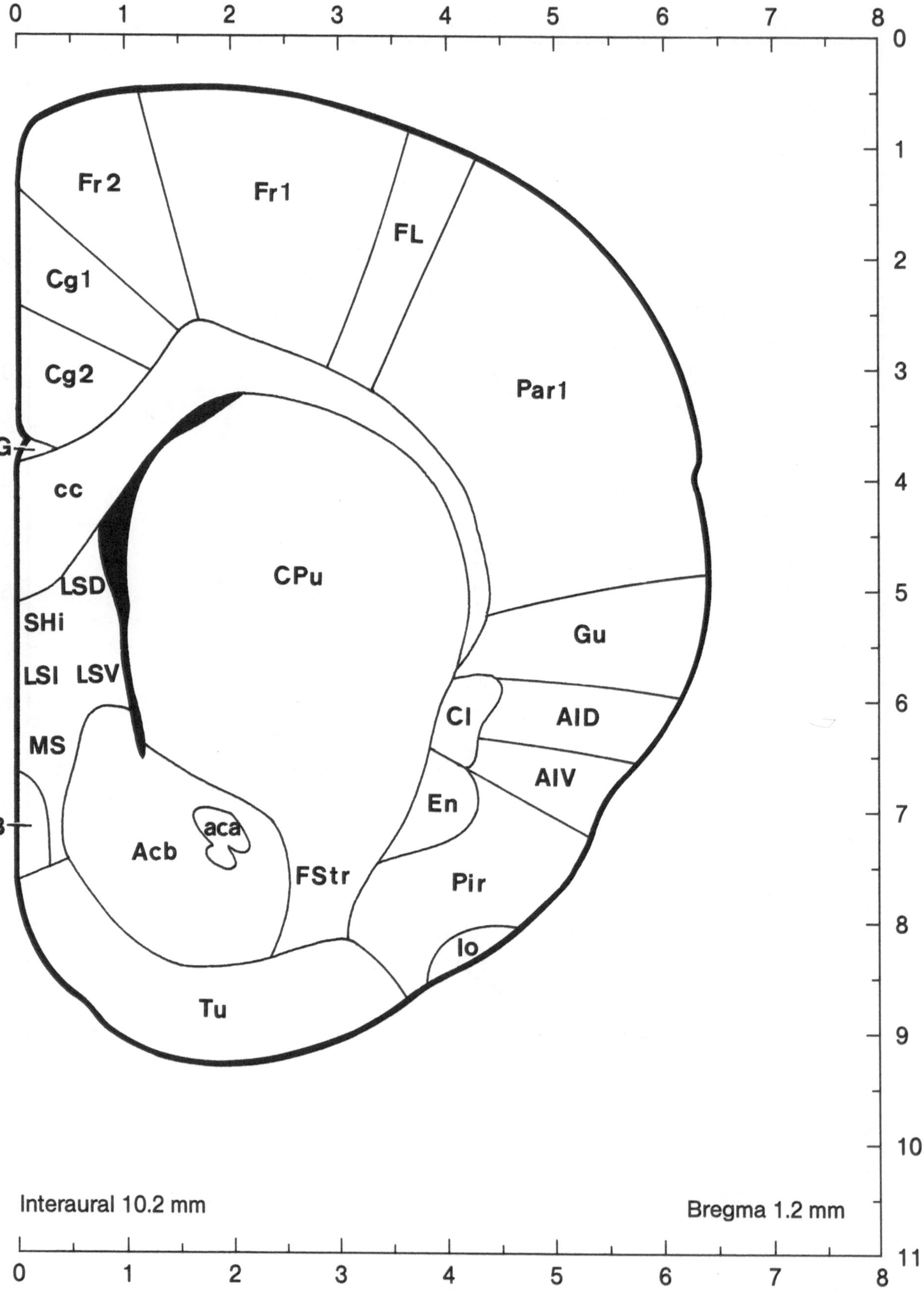

0
1
2
3
4
5
6
7
8
Fr 2
Fr1
FL
Cg1
Cg2
Par1
IG
cc
CPu
LSD
SHi
LSI
LSV
Gu
Cl
AID
MS
AIV
En
VDB
aca
Acb
FStr
Pir
lo
Tu
9
10
11
Interaural 10.2 mm
Bregma 1.2 mm

[18]

Fig. 10

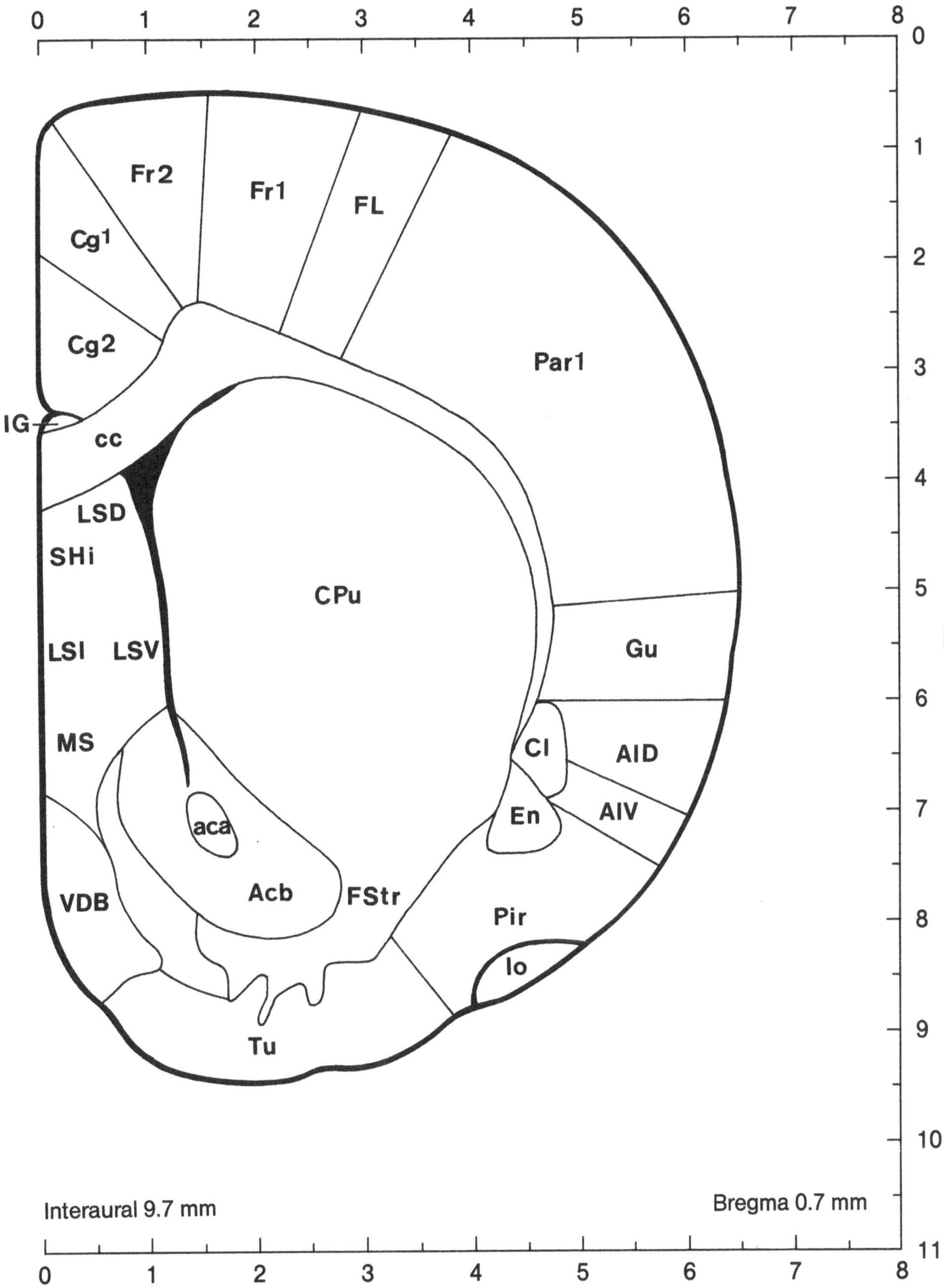

0 1 2 3 4 5 6 7 8
0 1 2 3 4 5 6 7 8 9 10 11
Fr2
Fr1
FL
Cg1
Cg2
Par1
IG
cc
LSD
SHi
CPu
LSI
LSV
Gu
MS
Cl
AID
En
AIV
aca
VDB
Acb
FStr
Pir
lo
Tu
Interaural 9.7 mm
Bregma 0.7 mm
0 1 2 3 4 5 6 7 8

Fig. 11

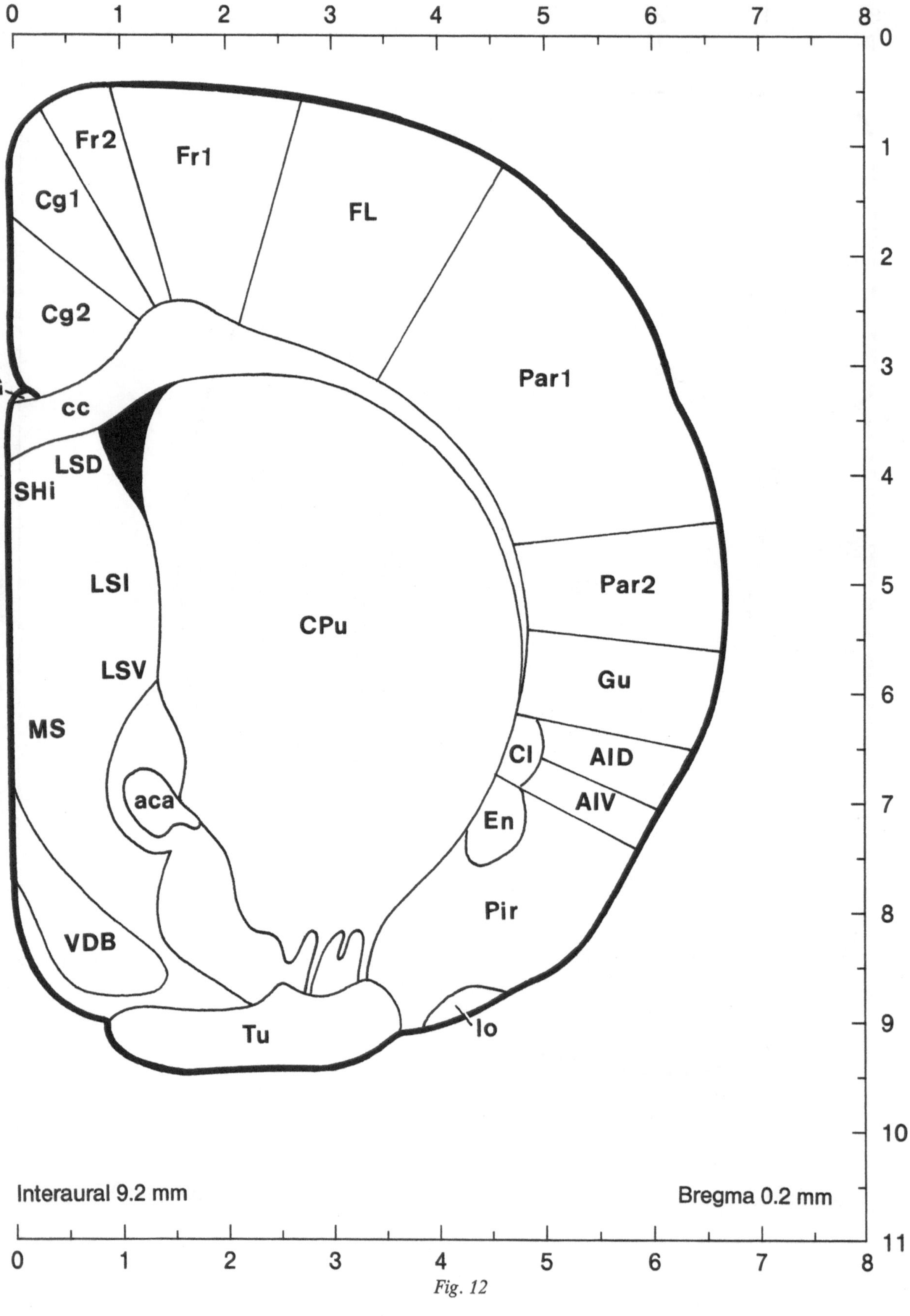
0
1
2
3
4
5
6
7
8
0
1
2
3
4
5
6
7
8
9
10
11
Fr2
Fr1
FL
Cg1
Cg2
Par1
IG
cc
LSD
SHi
LSI
CPu
Par2
LSV
Gu
MS
Cl
AID
aca
AIV
En
Pir
VDB
Tu
lo
Interaural 9.2 mm
Bregma 0.2 mm
0
1
2
3
4
5
6
7
8

[20]

Fig. 12

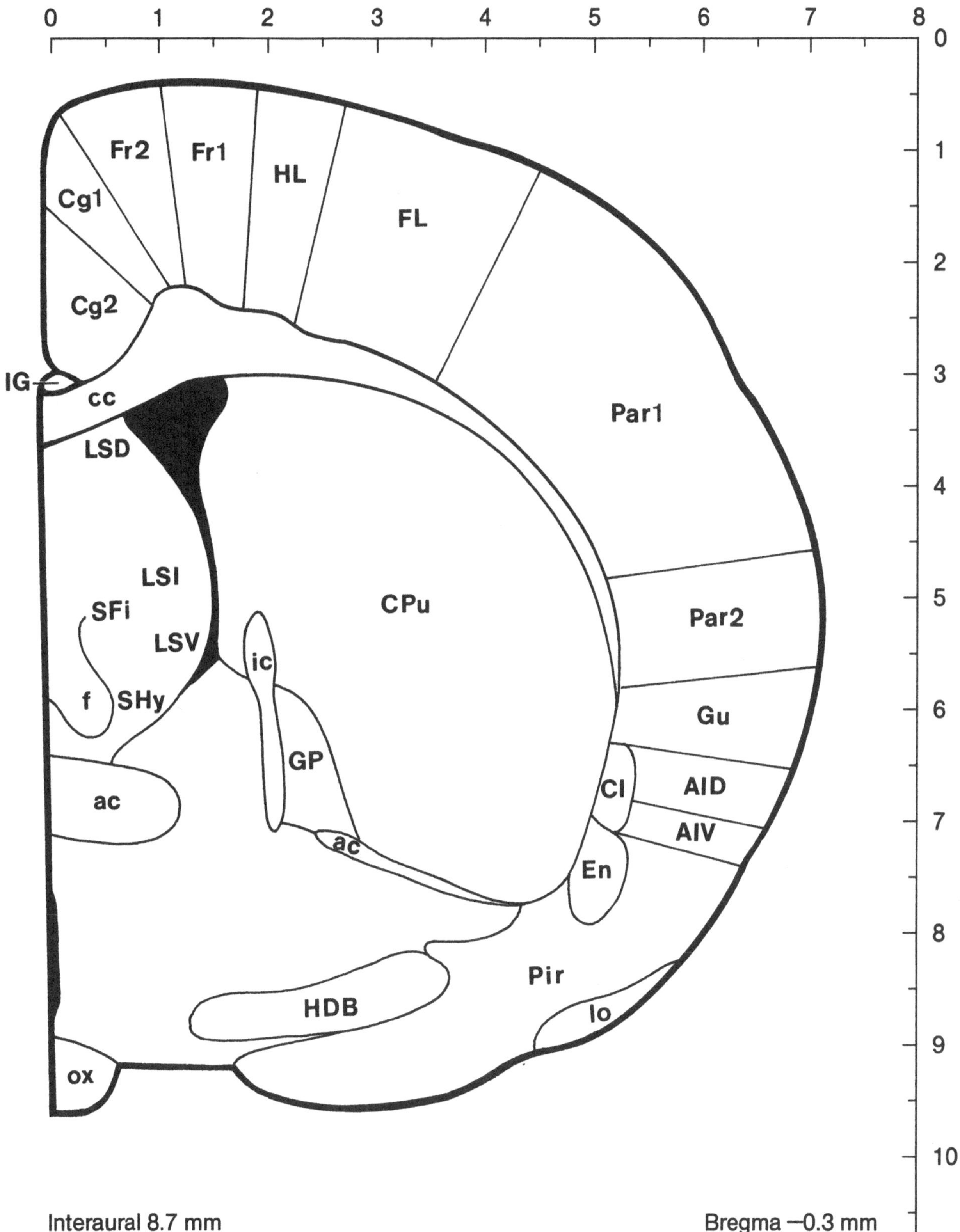

Fr2
Fr1
HL
FL
Cg1
Cg2
IG
cc
LSD
Par1
LSI
SFi
LSV
CPu
ic
Par2
Gu
f
SHy
GP
ac
Cl
AID
AIV
ac
En
Pir
HDB
lo
ox
Interaural 8.7 mm
Bregma −0.3 mm

[21]

Fig. 13

[22]

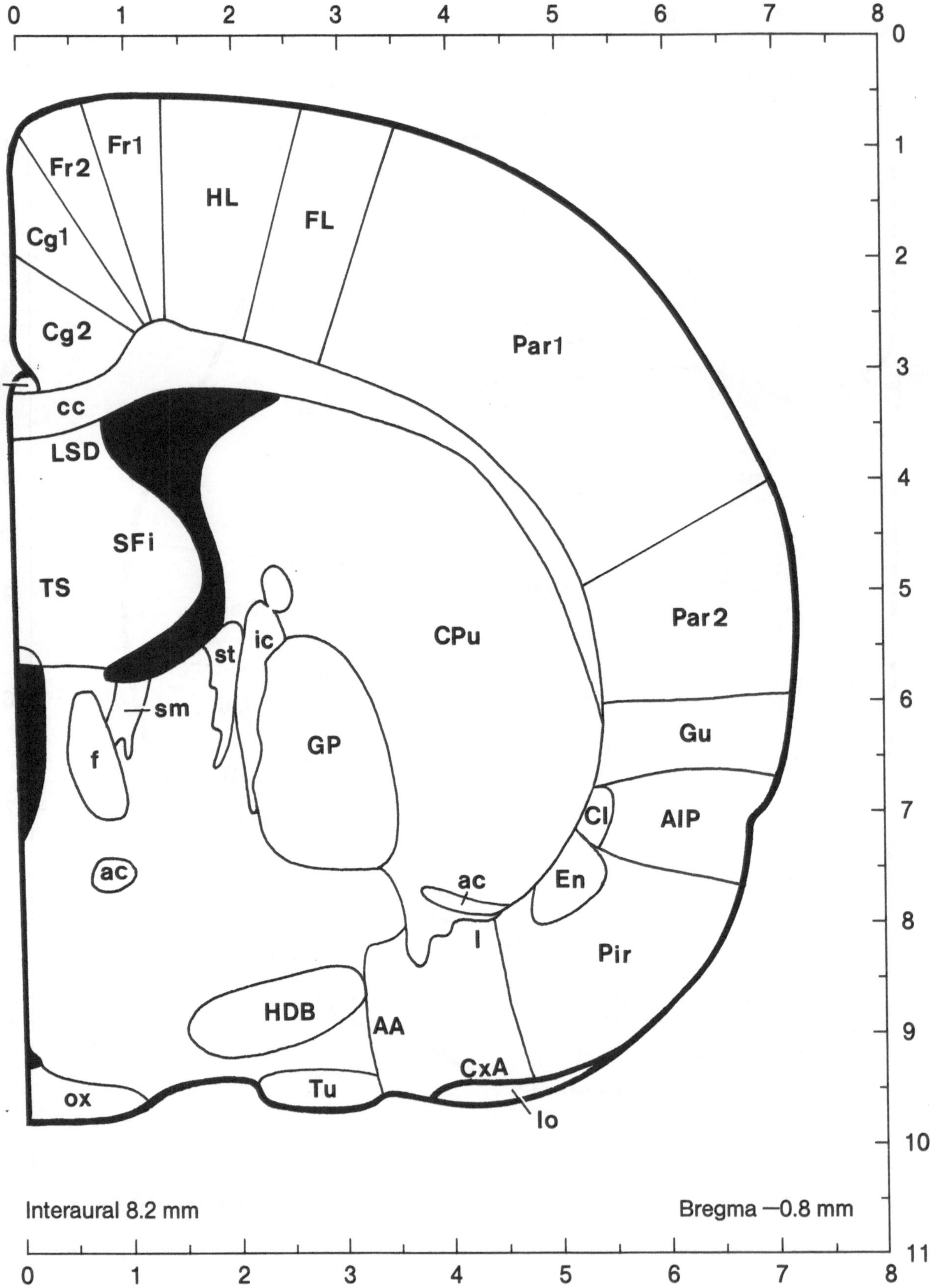
Fr1
Fr2
HL
FL
Cg1
Cg2
Par1
IG
cc
LSD
SFi
TS
Par2
ic
CPu
st
sm
GP
Gu
f
Cl
AIP
ac
ac
En
l
Pir
HDB
AA
CxA
ox
Tu
lo
Interaural 8.2 mm
Bregma —0.8 mm

Fig. 14

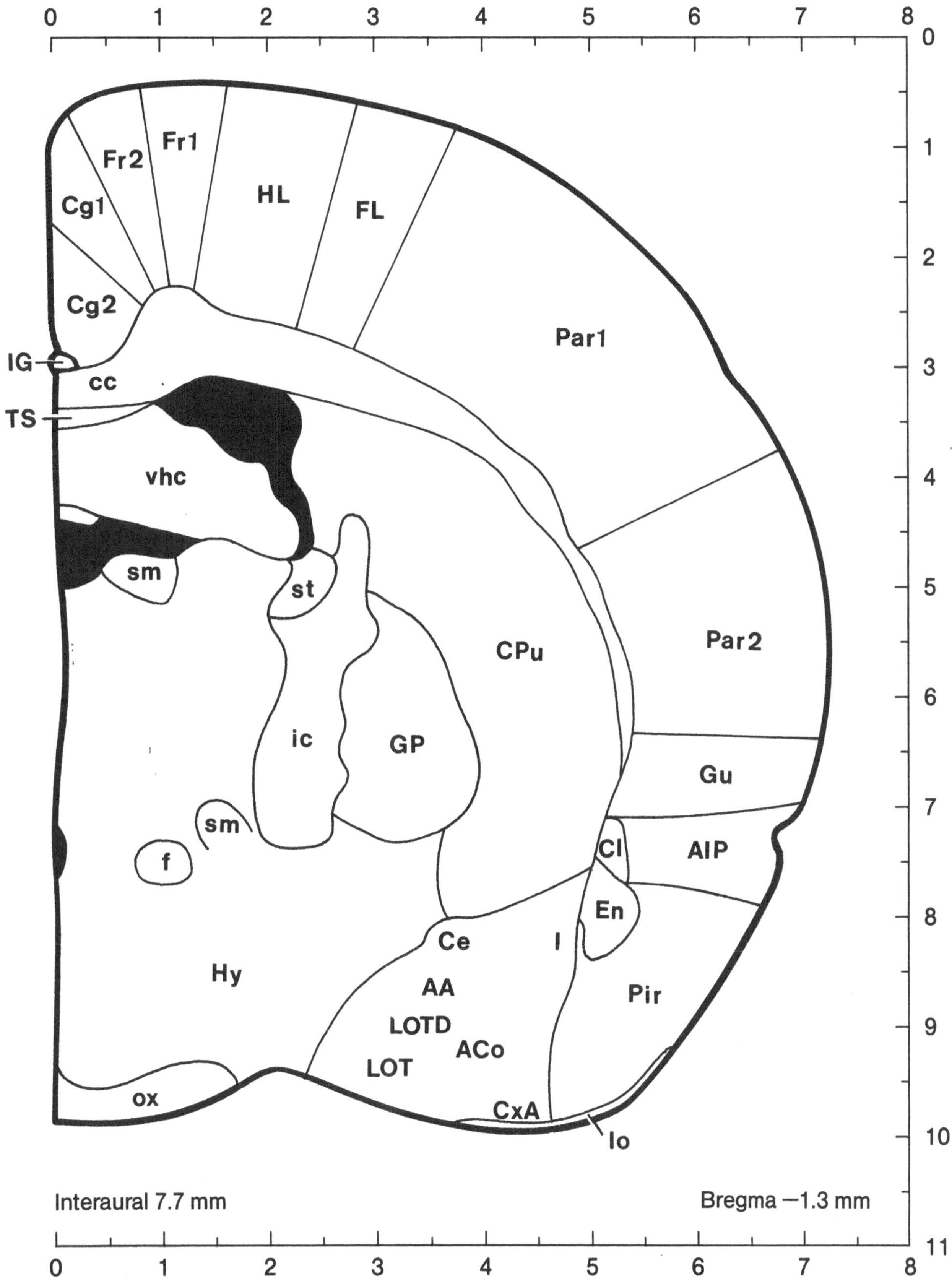
0 1 2 3 4 5 6 7 8
0 1 2 3 4 5 6 7 8 9 10 11
Fr1
Fr2
Cg1
HL
FL
Cg2
Par1
IG
cc
TS
vhc
sm
st
CPu
Par2
ic
GP
Gu
sm
Cl
AIP
f
En
Ce
I
Hy
AA
Pir
LOTD
ACo
LOT
ox
CxA
lo
Interaural 7.7 mm
Bregma −1.3 mm

Fig. 15

[24]

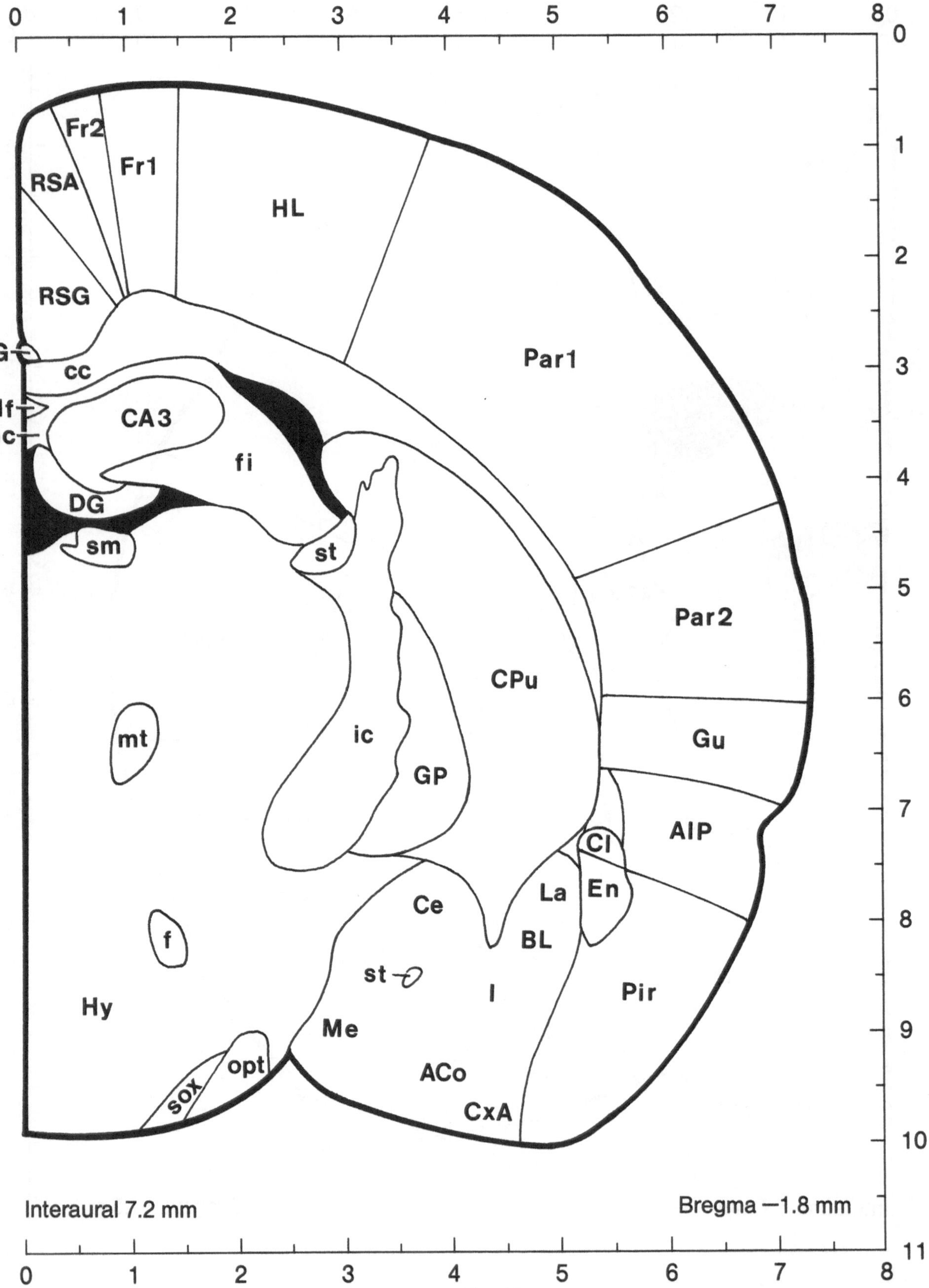

Fr2
Fr1
RSA
HL
RSG
Par1
IG
cc
df
dhc
CA3
fi
DG
sm
st
Par2
CPu
Gu
mt
ic
GP
AIP
Cl
La
En
Ce
f
BL
st
I
Pir
Hy
Me
opt
sox
ACo
CxA
Interaural 7.2 mm
Bregma −1.8 mm

Fig. 16

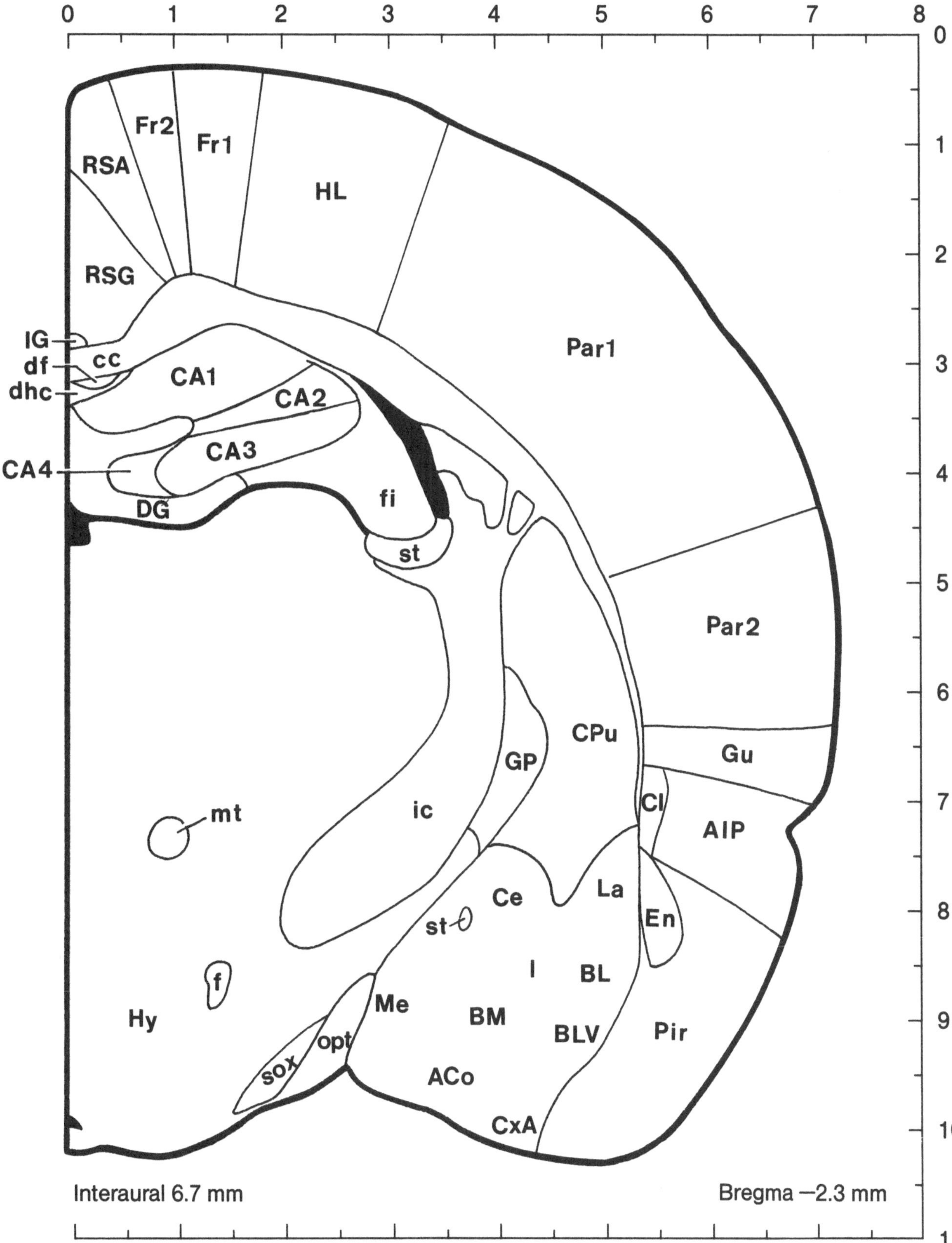
0 1 2 3 4 5 6 7 8
0 1 2 3 4 5 6 7 8 9 10 11
Fr2
Fr1
RSA
HL
RSG
Par1
IG
df
cc
dhc
CA1
CA2
CA3
CA4
fi
DG
st
Par2
CPu
Gu
GP
mt
ic
Cl
AIP
La
Ce
En
st
I
BL
f
Me
BM
Hy
BLV
Pir
opt
sox
ACo
CxA
Interaural 6.7 mm
Bregma −2.3 mm

Fig. 17

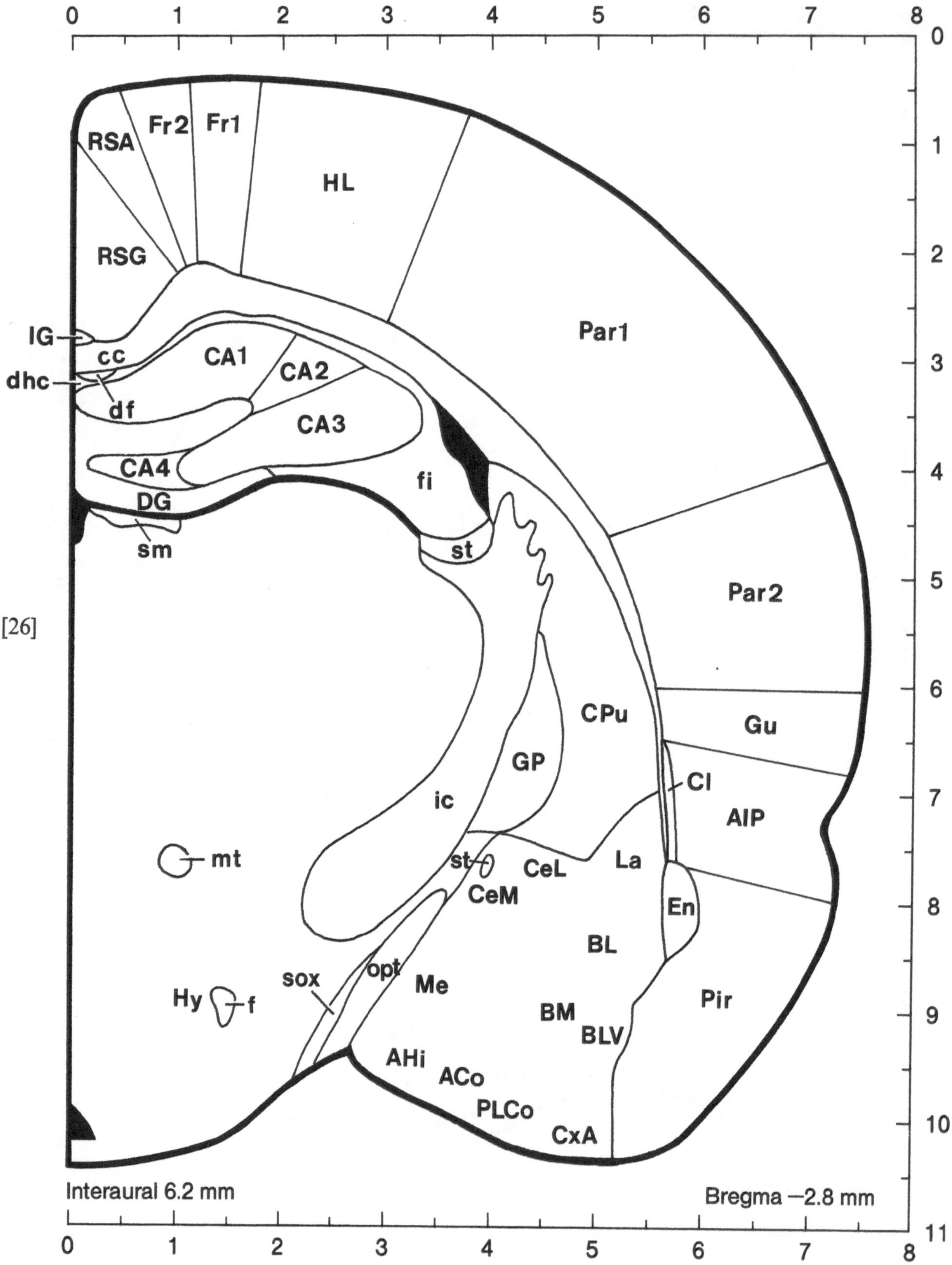

RSA
Fr2
Fr1
HL
RSG
Par1
IG
cc
CA1
CA2
dhc
df
CA3
CA4
fi
DG
sm
st
Par2
CPu
Gu
GP
Cl
ic
AIP
mt
st
CeL
La
CeM
En
BL
sox
opt
Me
Pir
Hy
f
BM
BLV
AHi
ACo
PLCo
CxA
Interaural 6.2 mm
Bregma −2.8 mm

[26]

Fig. 18

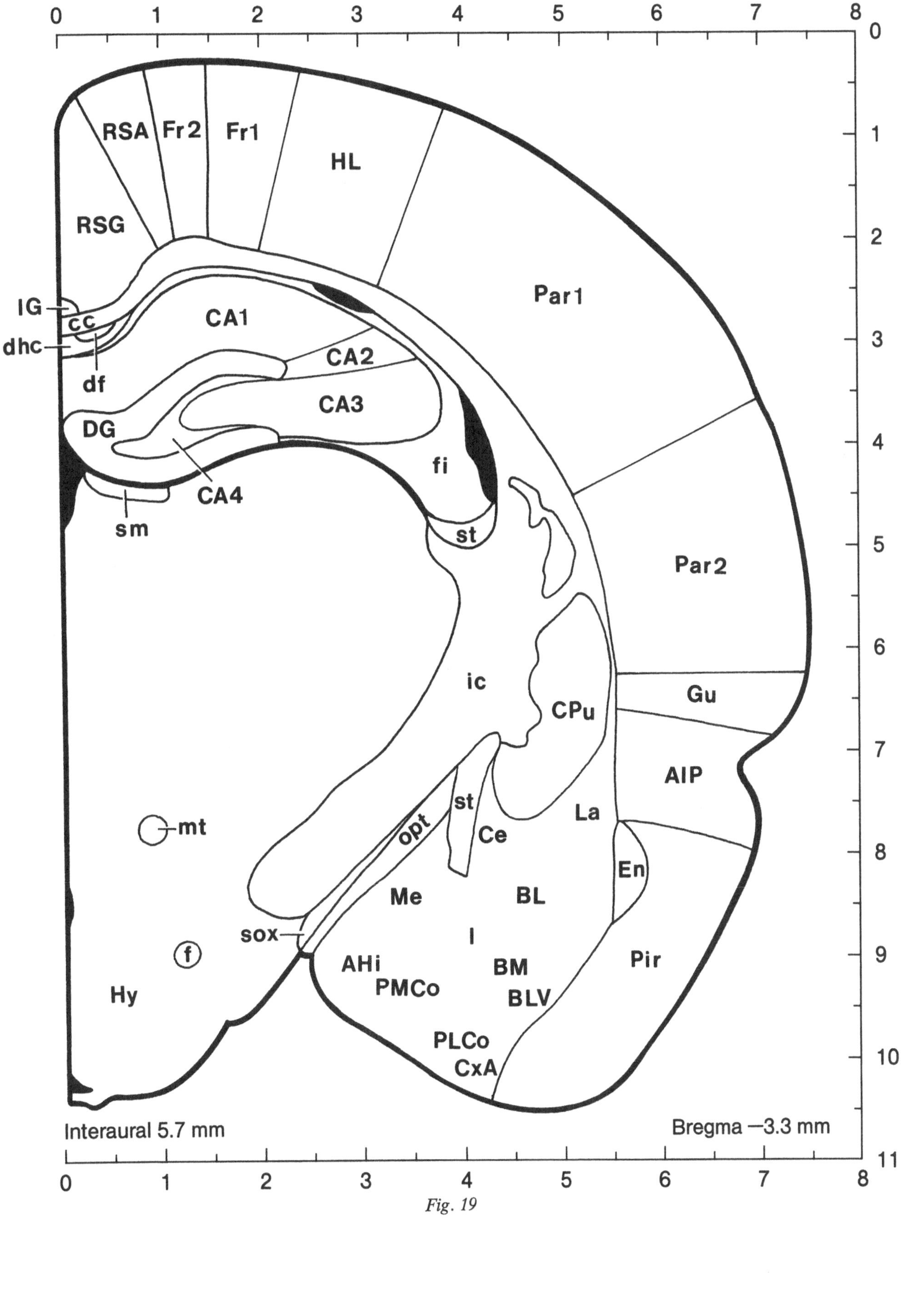

0
1
2
3
4
5
6
7
8
9
10
11
RSA
Fr2
Fr1
HL
RSG
Par1
IG
cc
CA1
dhc
CA2
df
CA3
DG
fi
CA4
sm
st
Par2
ic
Gu
CPu
AIP
st
La
mt
opt
Ce
En
Me
BL
sox
I
f
AHi
BM
Pir
PMCo
Hy
BLV
PLCo
CxA
Interaural 5.7 mm
Bregma −3.3 mm

Fig. 19

[28]

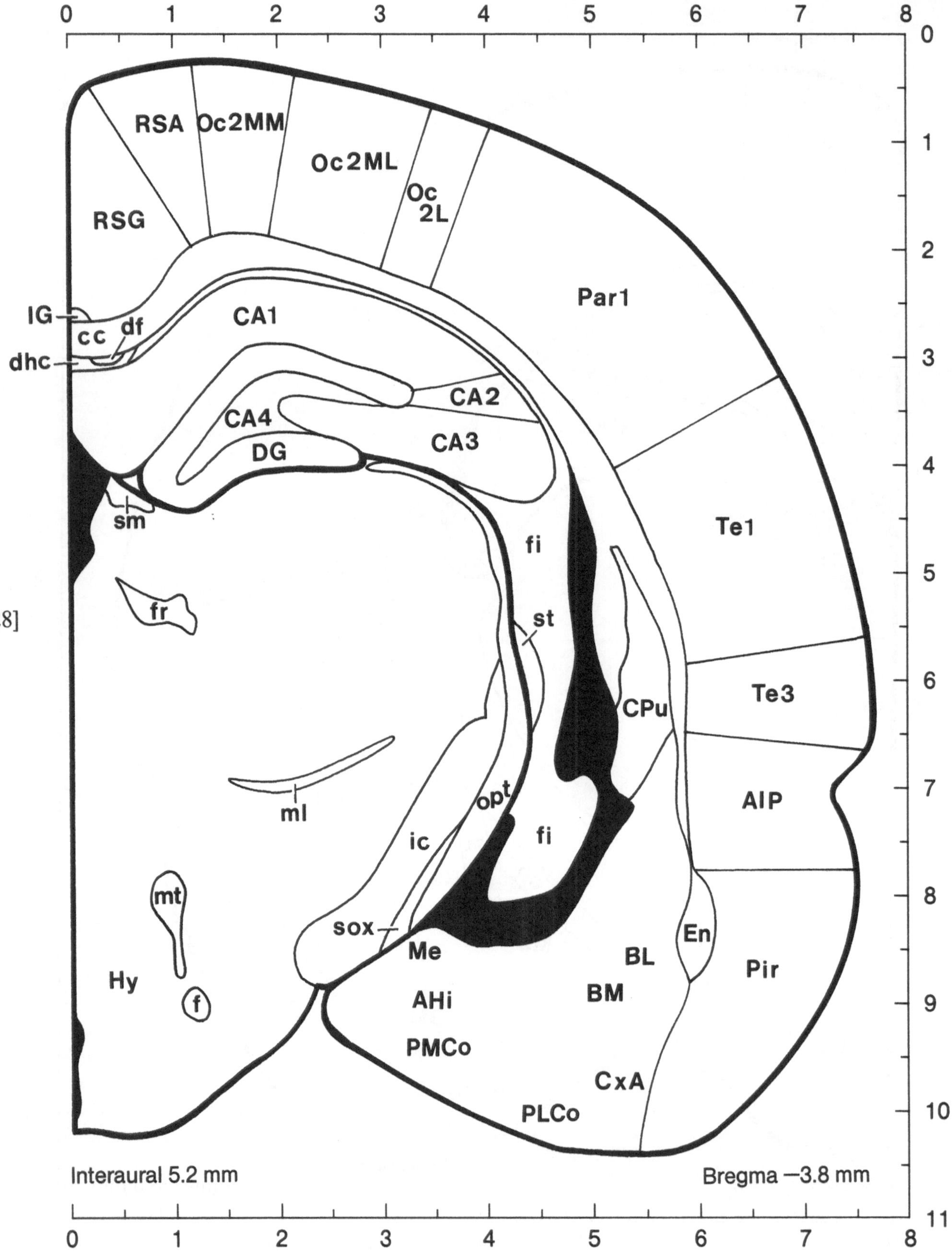
0 1 2 3 4 5 6 7 8
0 1 2 3 4 5 6 7 8 9 10 11
RSA
Oc2MM
Oc2ML
Oc
2L
RSG
Par1
IG
cc
df
dhc
CA1
CA2
CA4
CA3
DG
sm
fi
Te1
fr
st
Te3
CPu
AIP
ml
opt
fi
ic
mt
sox
Me
En
BL
Pir
Hy
AHi
BM
f
PMCo
CxA
PLCo
Interaural 5.2 mm
Bregma −3.8 mm

Fig. 20

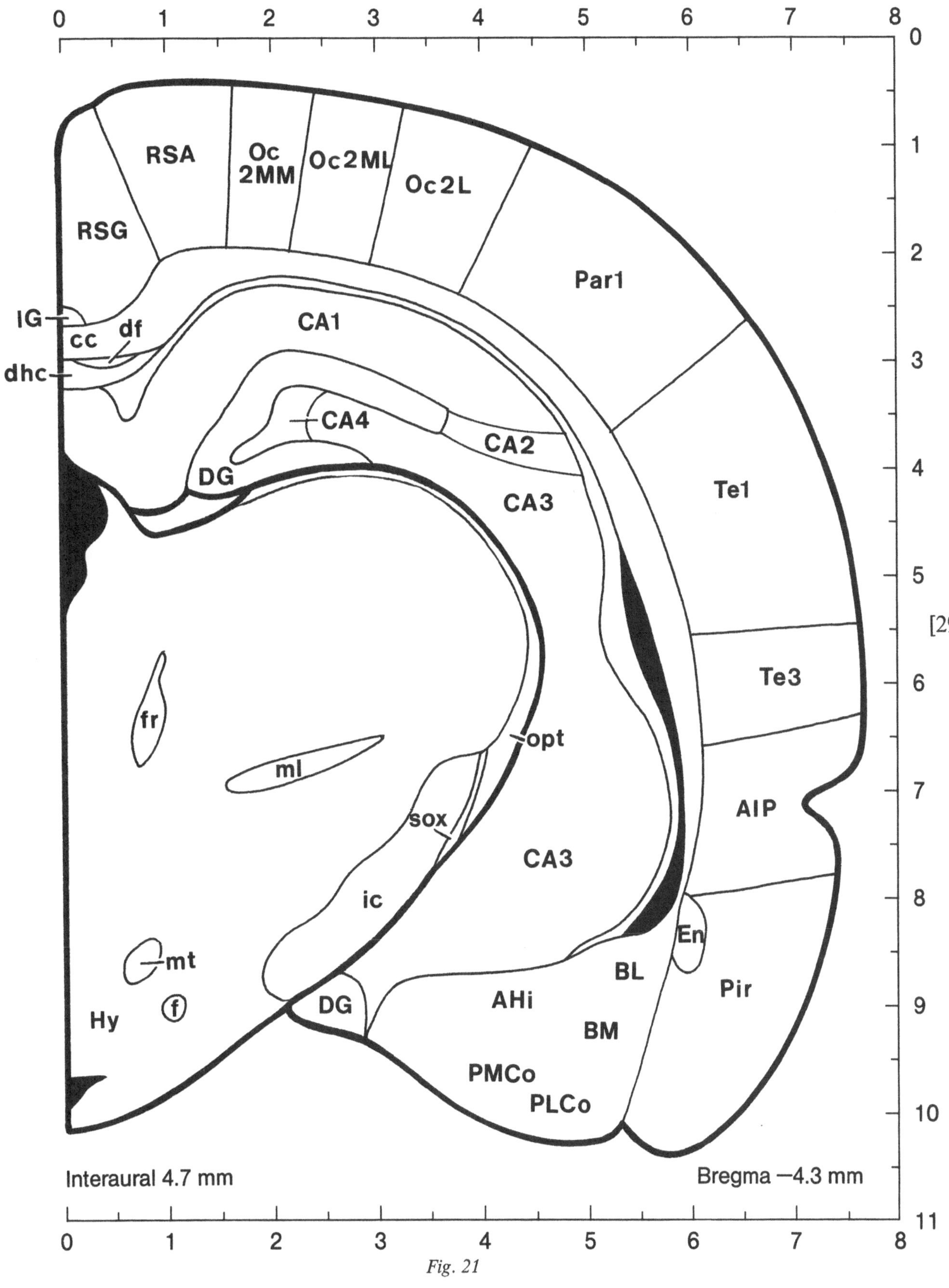

RSA
Oc 2MM
Oc2ML
Oc2L
RSG
Par1
IG
cc
df
CA1
dhc
CA4
CA2
DG
CA3
Te1
Te3
fr
opt
ml
AIP
sox
CA3
ic
En
mt
BL
Pir
DG
AHi
Hy
f
BM
PMCo
PLCo
Interaural 4.7 mm
Bregma −4.3 mm

[29]

Fig. 21

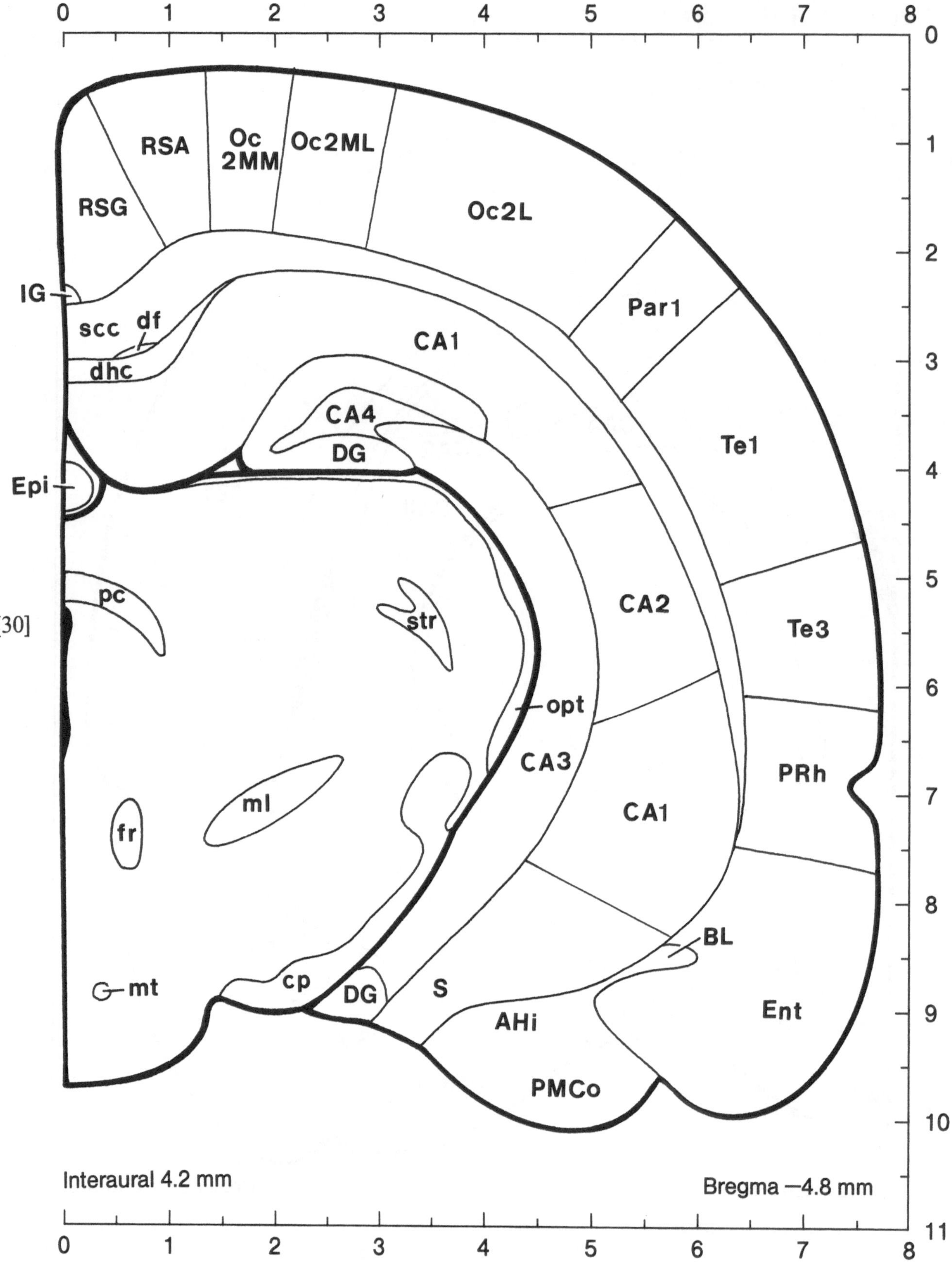

0
1
2
3
4
5
6
7
8
RSA
Oc
2MM
Oc2ML
RSG
Oc2L
Par1
IG
scc
df
dhc
CA1
CA4
DG
Te1
Epi
CA2
Te3
pc
str
opt
CA3
PRh
ml
CA1
fr
BL
mt
cp
DG
S
Ent
AHi
PMCo
9
10
11
Interaural 4.2 mm
Bregma −4.8 mm

[30]

Fig. 22

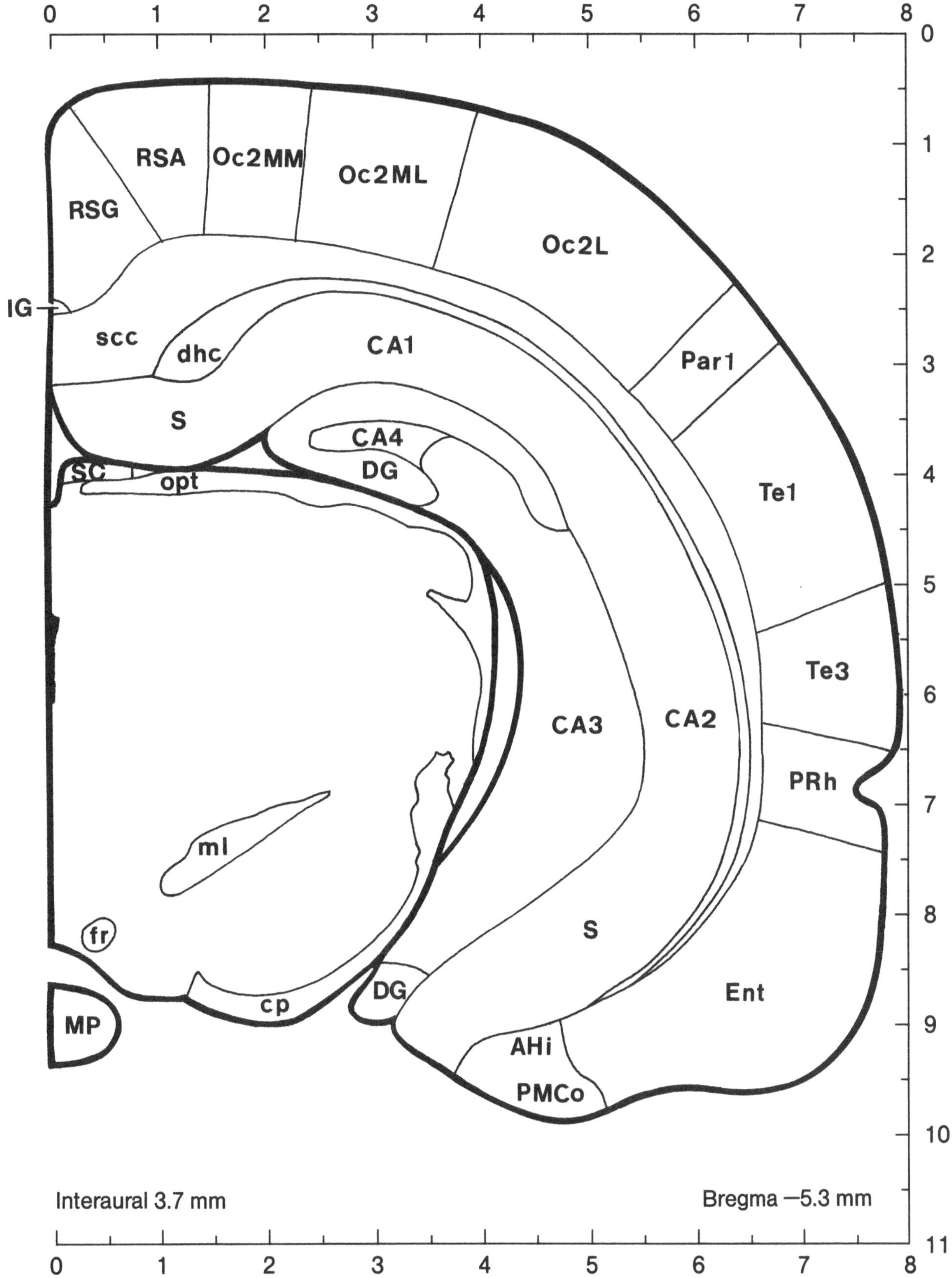
0
1
2
3
4
5
6
7
8
RSA
Oc2MM
Oc2ML
RSG
Oc2L
IG
scc
dhc
CA1
Par1
S
CA4
DG
SC
opt
Te1
Te3
CA3
CA2
PRh
ml
S
fr
DG
Ent
cp
MP
AHi
PMCo
9
10
11
Interaural 3.7 mm
Bregma −5.3 mm

Fig. 23

[32]

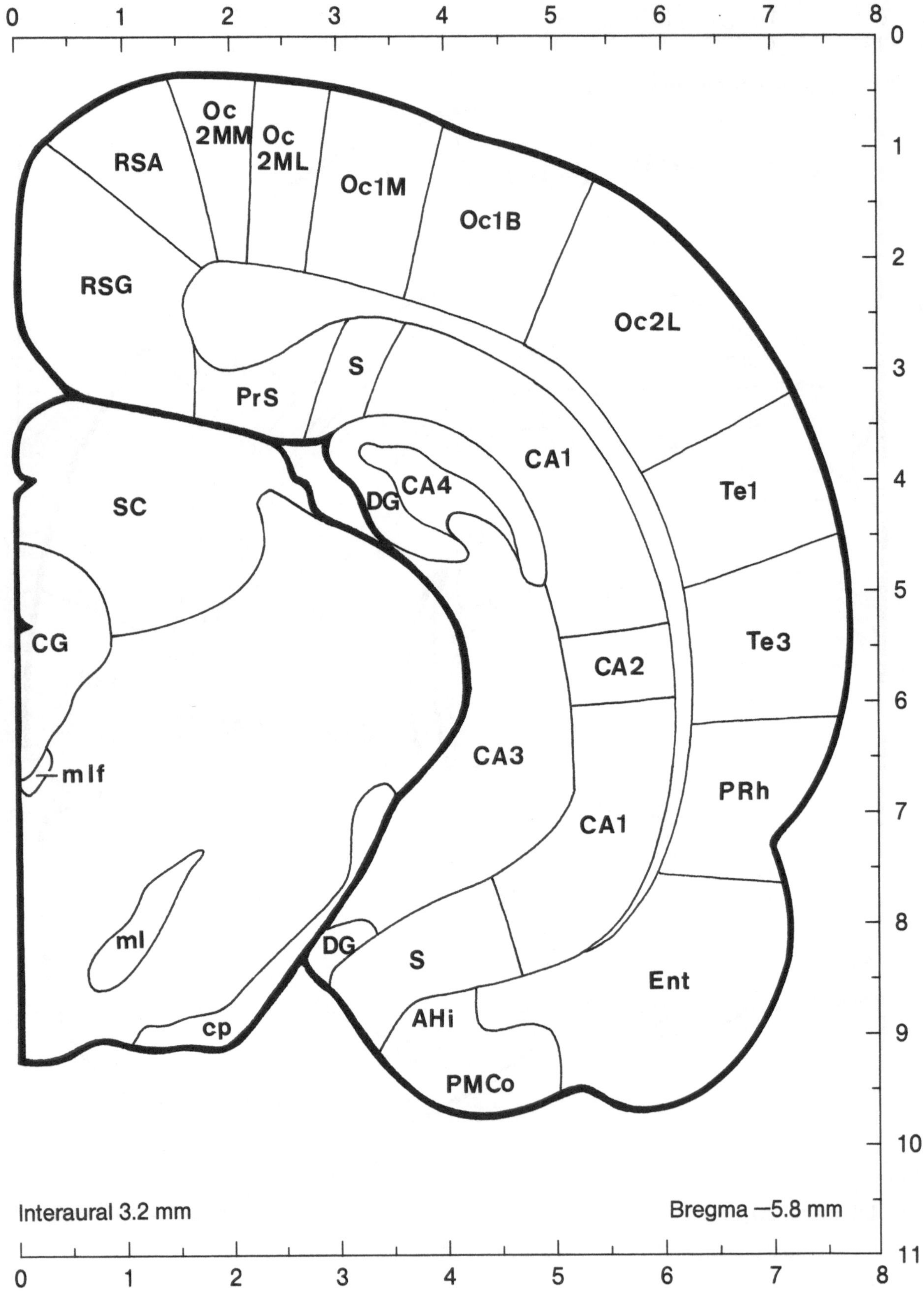
Oc
2MM
Oc
2ML
RSA
Oc1M
Oc1B
RSG
Oc2L
S
PrS
CA1
SC
CA4
DG
Te1
CG
Te3
CA2
CA3
mlf
PRh
CA1
DG
ml
S
Ent
AHi
cp
PMCo
Interaural 3.2 mm
Bregma −5.8 mm

Fig. 24

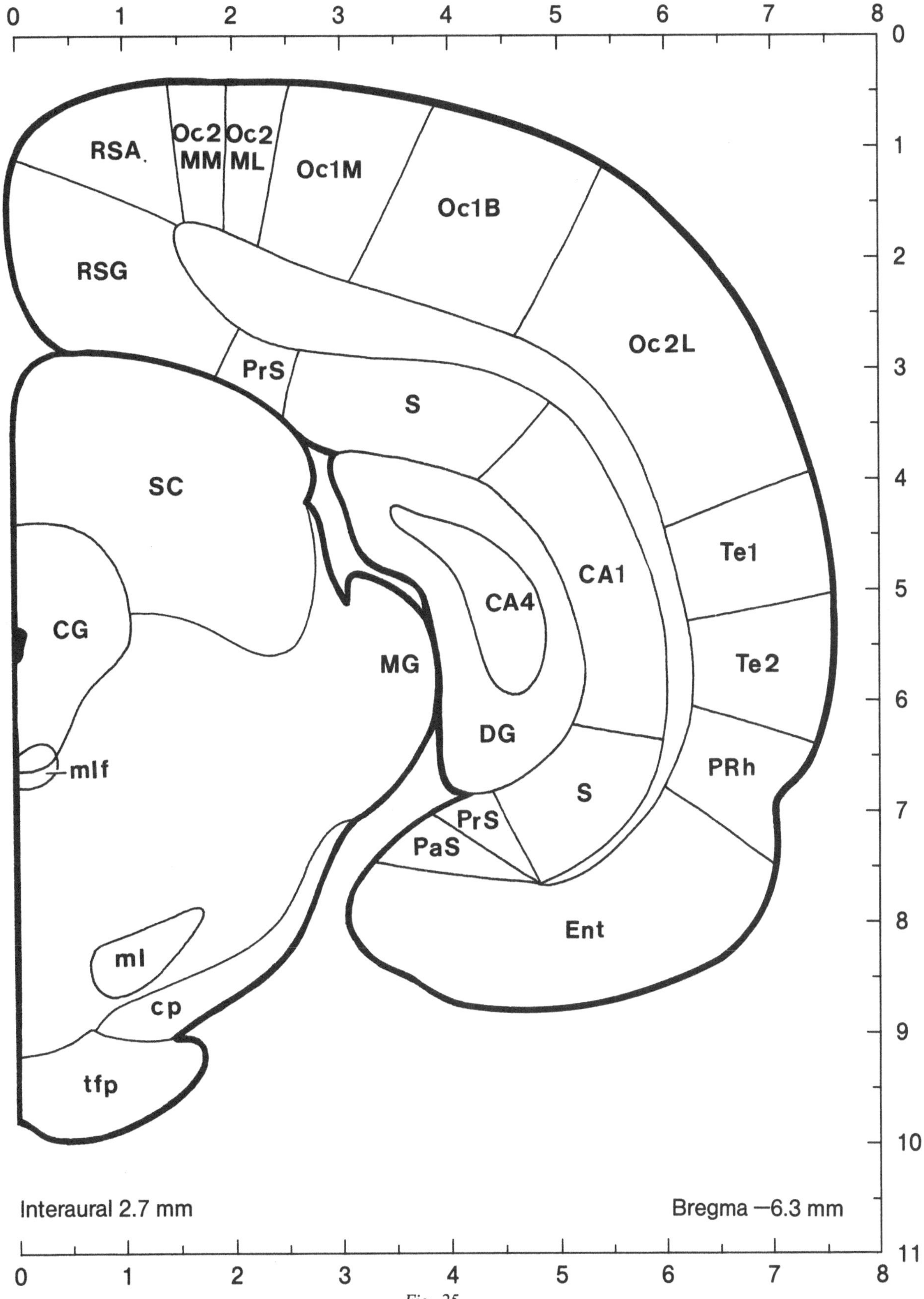

0
1
2
3
4
5
6
7
8
RSA
Oc2 MM
Oc2 ML
Oc1M
Oc1B
RSG
Oc2L
PrS
S
SC
Te1
CA1
CA4
CG
MG
Te2
DG
mlf
PRh
S
PrS
PaS
Ent
ml
cp
tfp
9
10
11
Interaural 2.7 mm
Bregma −6.3 mm

[33]

Fig. 25

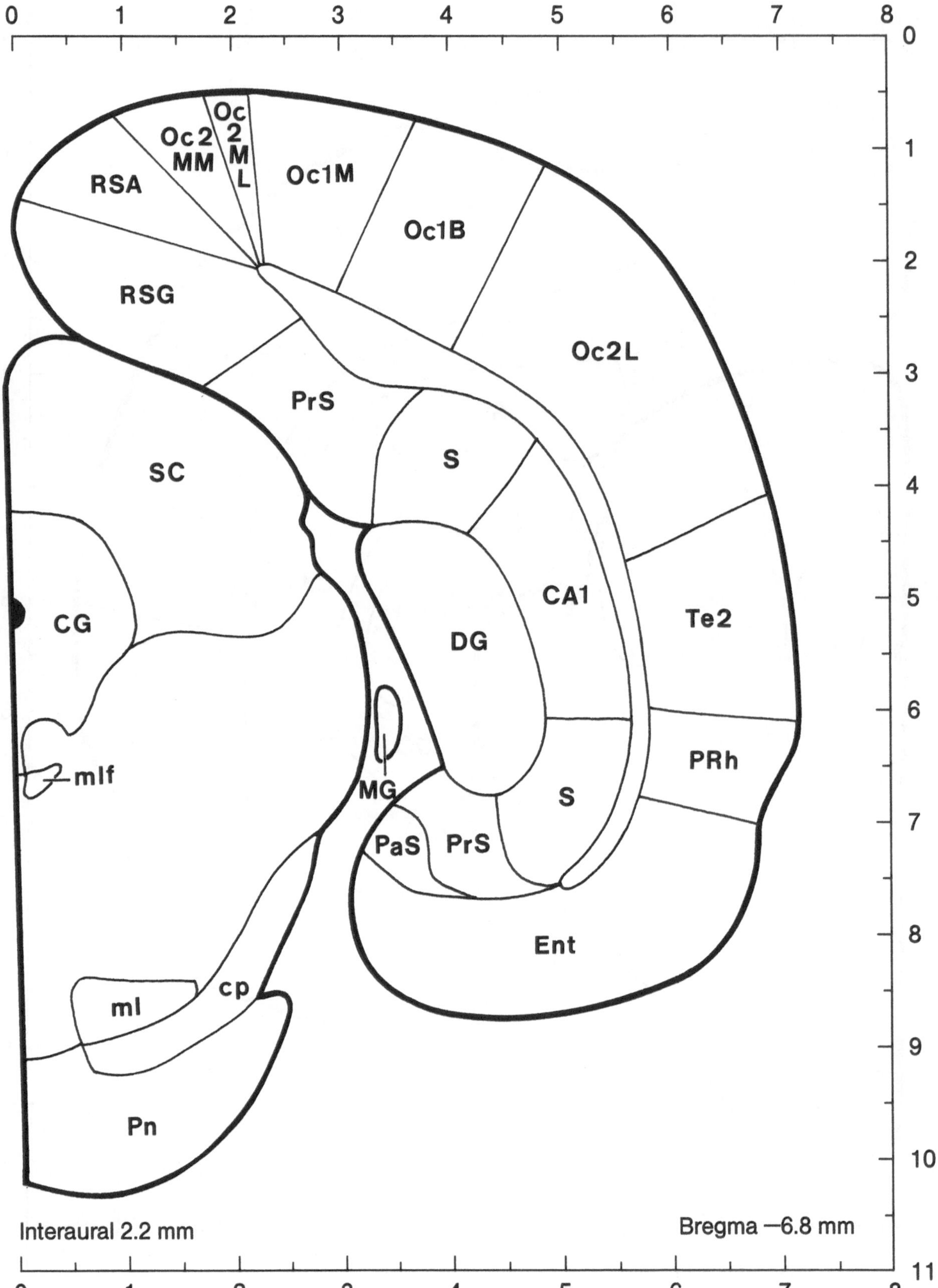
0
1
2
3
4
5
6
7
8
9
10
11
RSA
Oc2
MM
Oc
2
M
L
Oc1M
Oc1B
RSG
Oc2L
PrS
SC
S
CA1
CG
DG
Te2
mlf
MG
PRh
S
PaS
PrS
Ent
ml
cp
Pn
Interaural 2.2 mm
Bregma −6.8 mm

[34]

Fig. 26

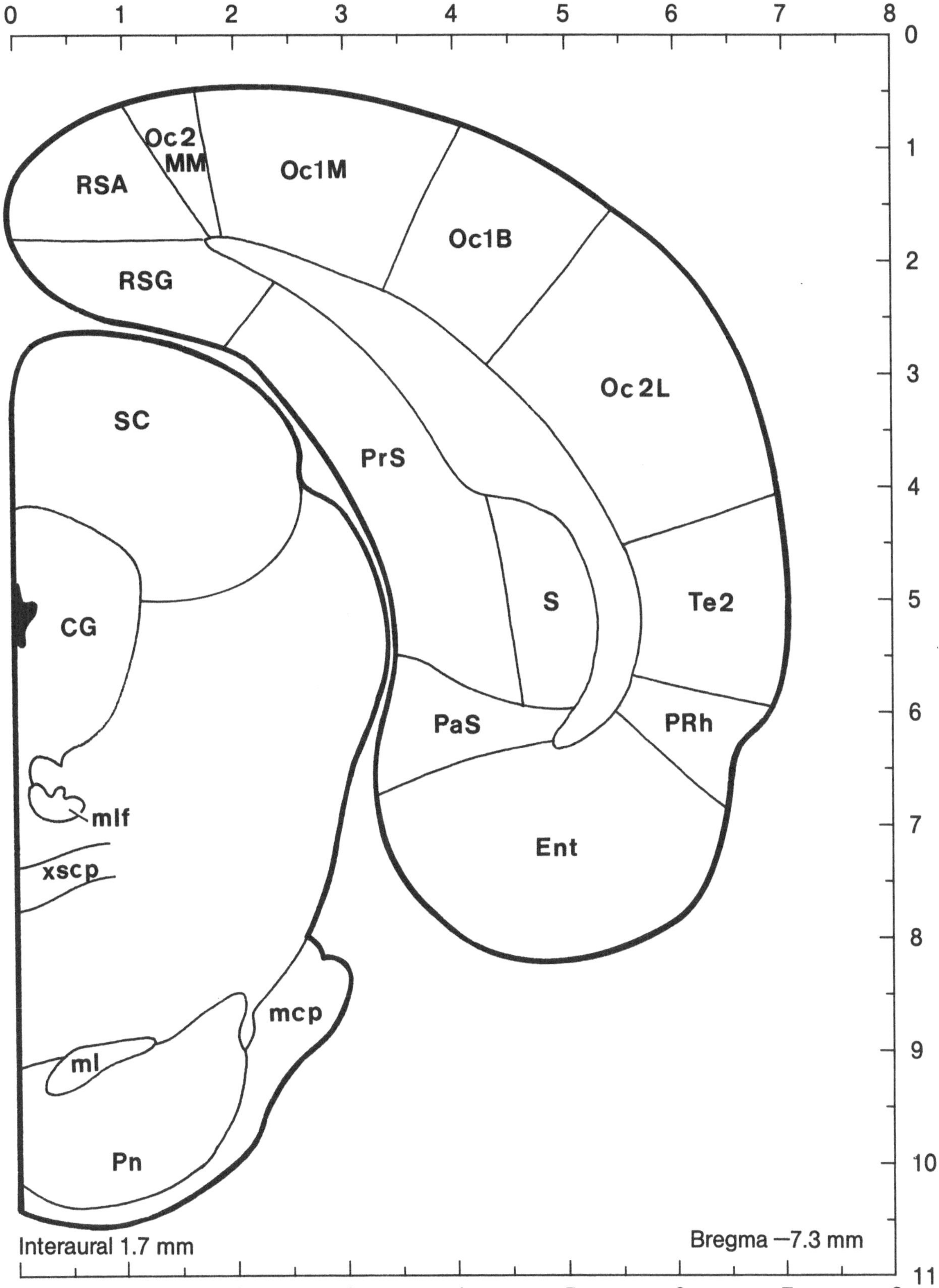
0 1 2 3 4 5 6 7 8
0 1 2 3 4 5 6 7 8 9 10 11
RSA
Oc2
MM
Oc1M
Oc1B
RSG
Oc 2L
SC
PrS
S
Te2
CG
PaS
PRh
mlf
Ent
xscp
mcp
ml
Pn
Interaural 1.7 mm
Bregma −7.3 mm
0 1 2 3 4 5 6 7 8

Fig. 27

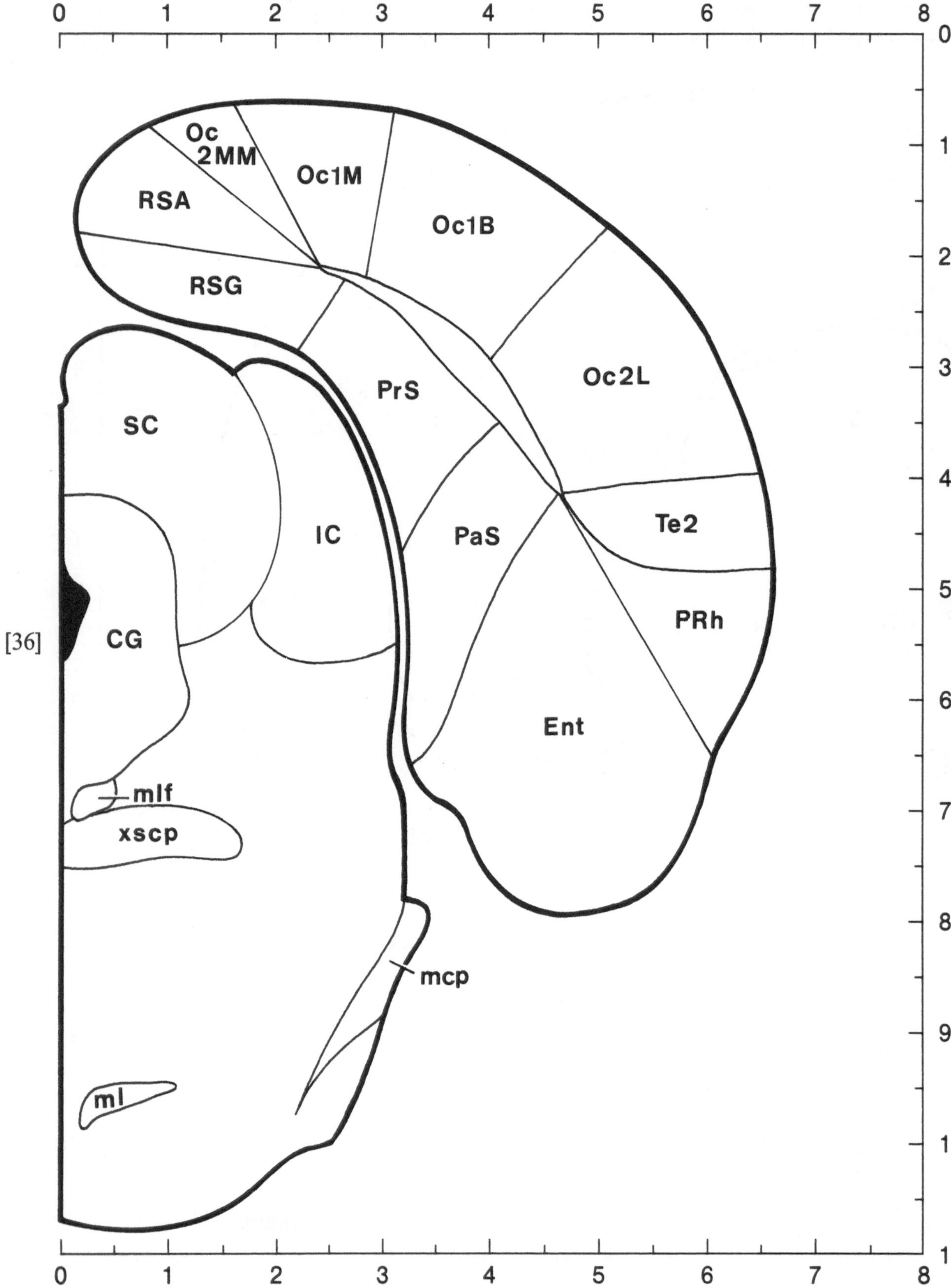

Interaural 1.2 mm *Fig. 28* Bregma −7.8 mm

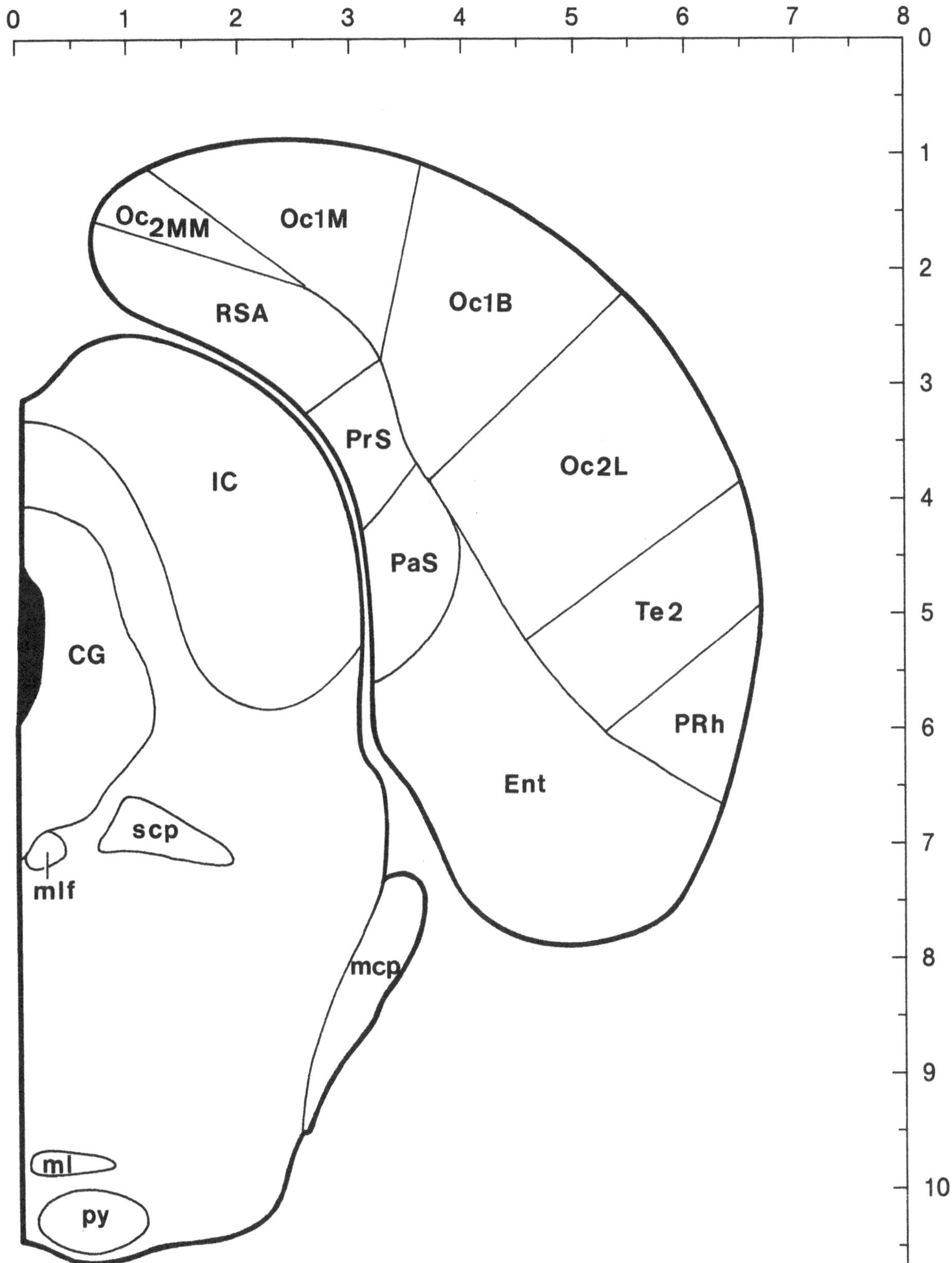

Interaural 0.7 mm

Fig. 29

Bregma −8.3 mm

[38]

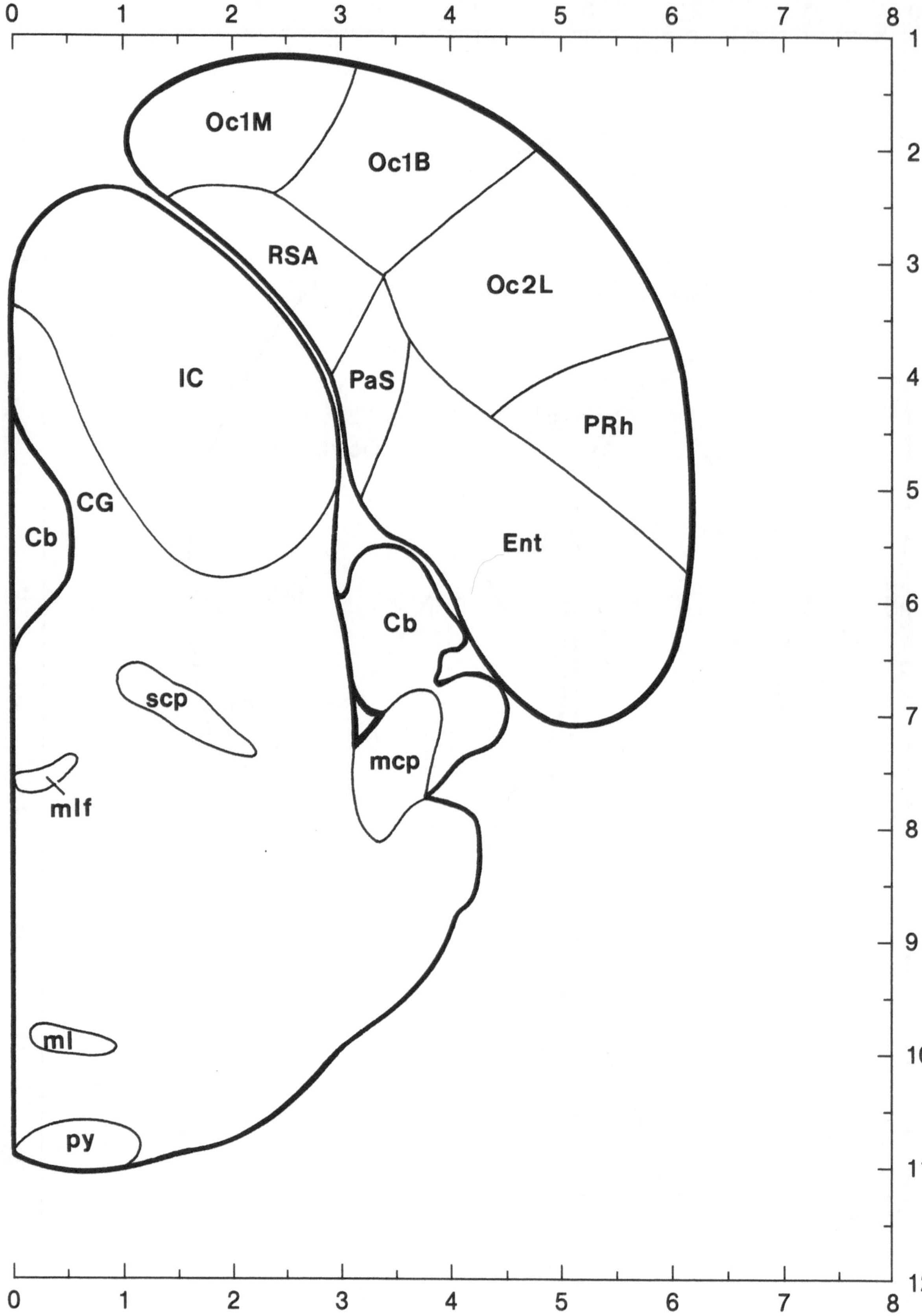

Interaural 0.2 mm

Fig. 30

Bregma −8.8 mm

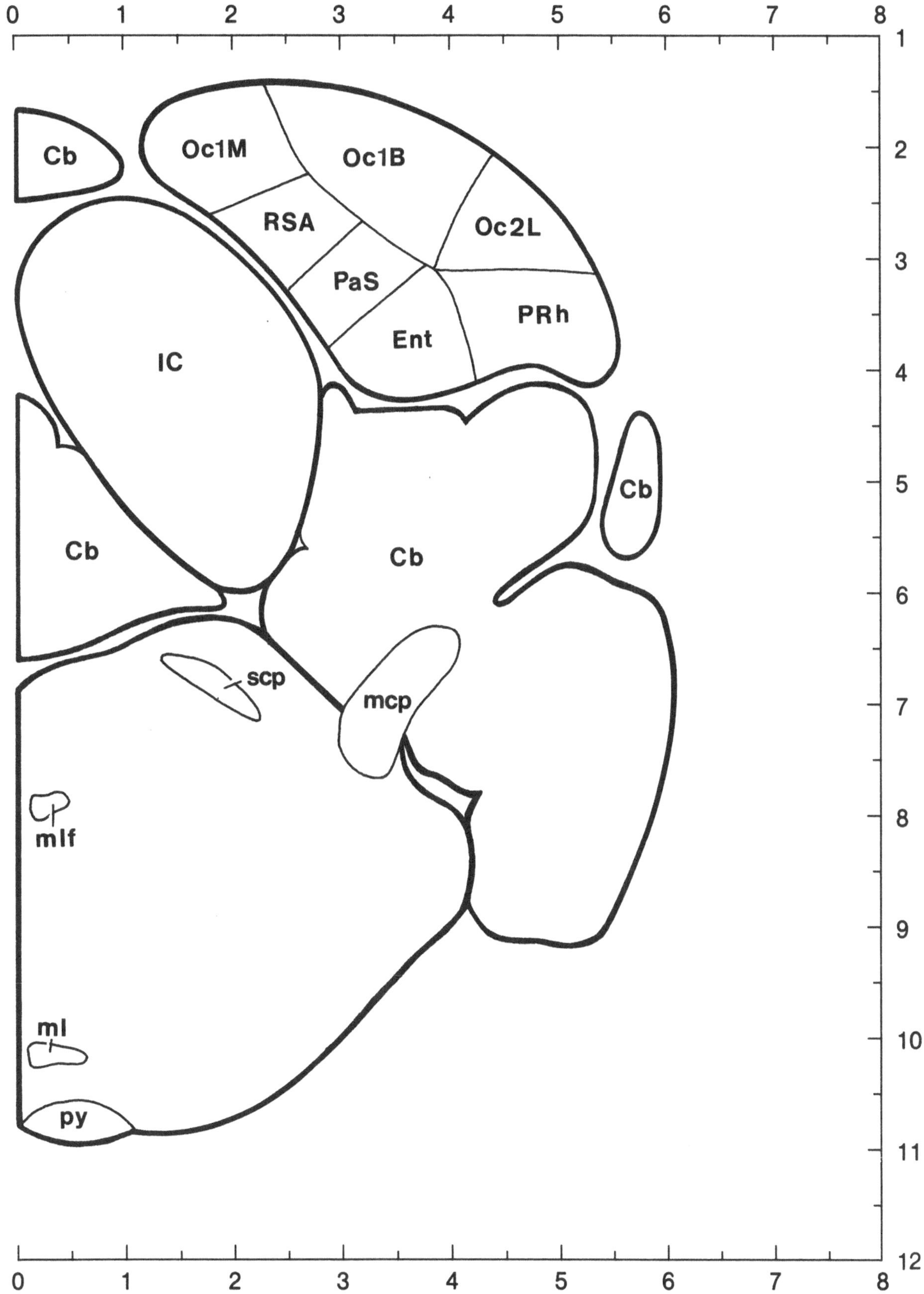

Interaural −0.3 mm

Fig. 31

Bregma −9.3 mm

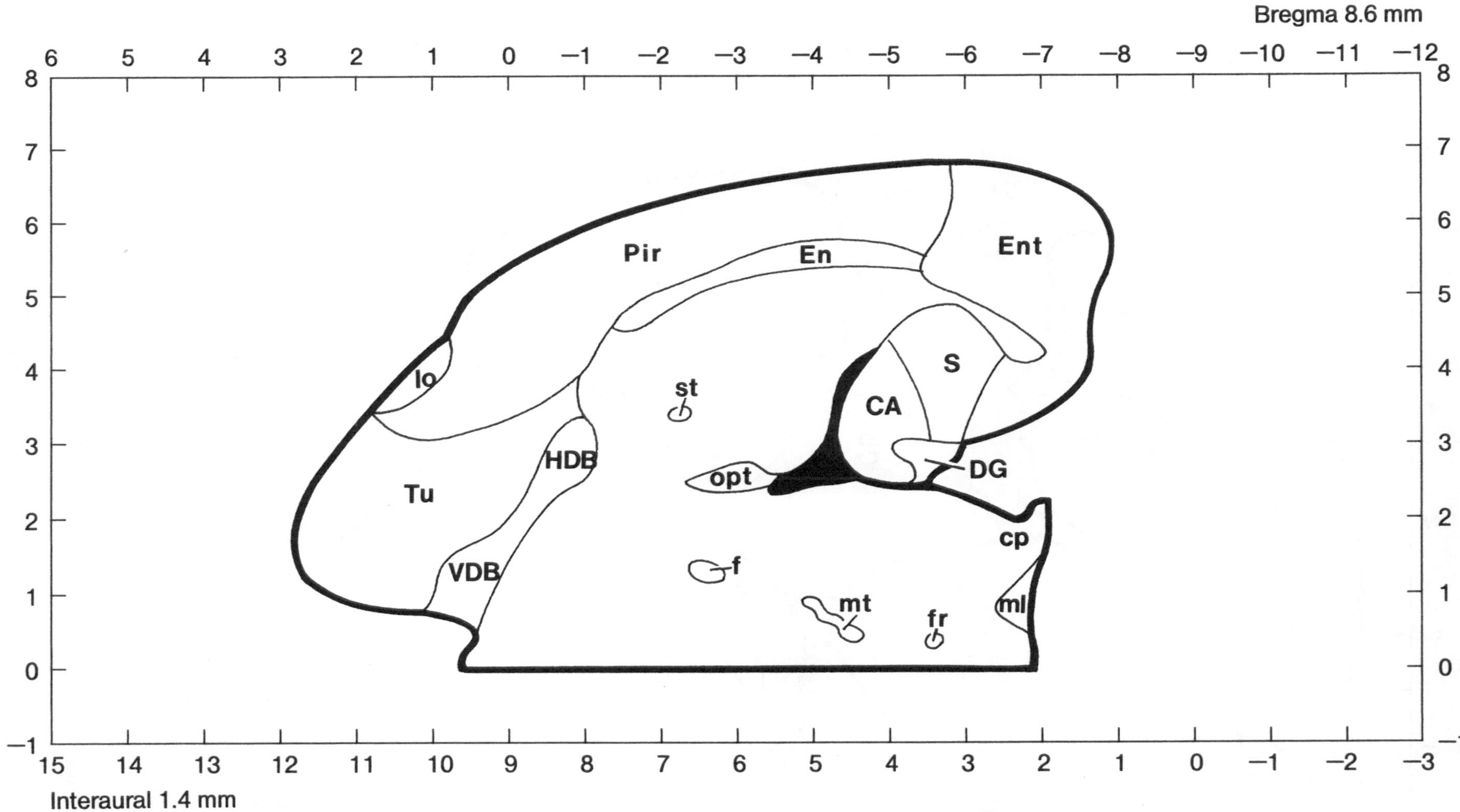
Bregma 8.6 mm
6
5
4
3
2
1
0
−1
−2
−3
−4
−5
−6
−7
−8
−9
−10
−11
−12
8
7
6
5
4
3
2
1
0
−1
Pir
En
Ent
S
lo
st
CA
HDB
opt
DG
Tu
cp
VDB
f
mt
fr
ml
15
14
13
12
11
10
9
8
7
6
5
4
3
2
1
0
−1
−2
−3
Interaural 1.4 mm

Fig. 32

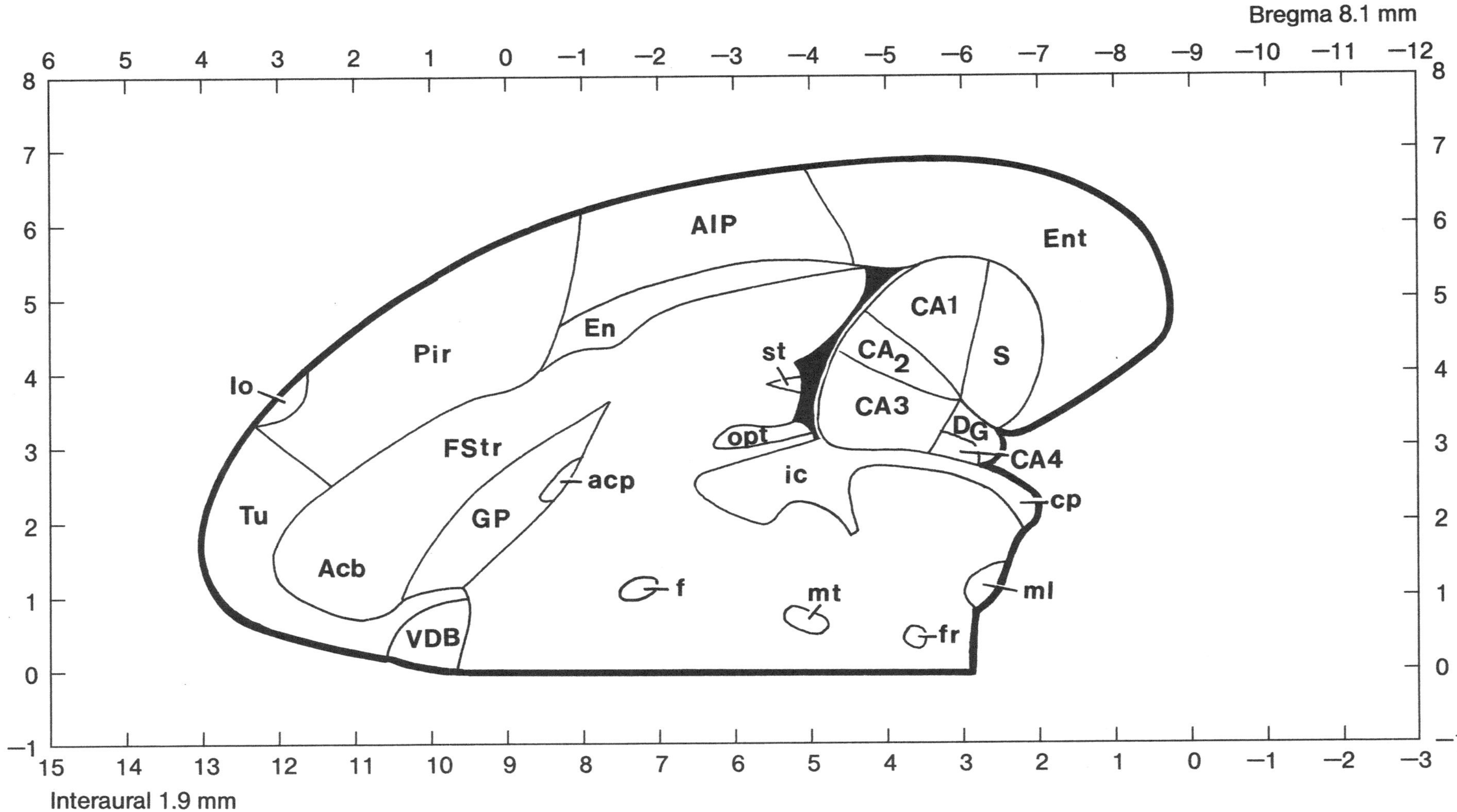

Bregma 8.1 mm
Interaural 1.9 mm
Ent
AIP
CA1
S
CA2
CA3
DG
CA4
cp
ml
fr
mt
f
ic
opt
st
En
Pir
lo
FStr
acp
GP
Tu
Acb
VDB

Fig. 33

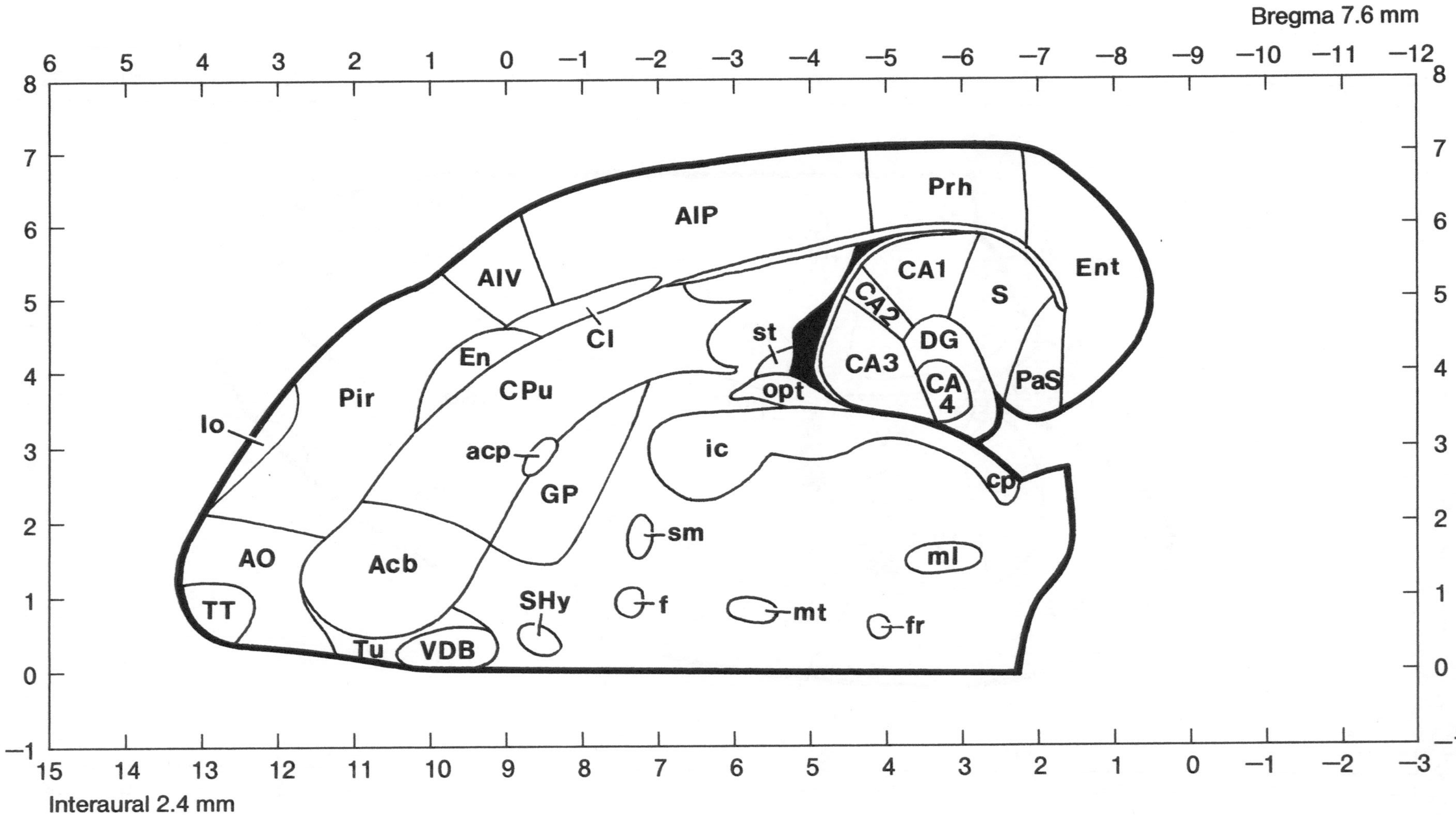
Bregma 7.6 mm
Interaural 2.4 mm
AIP
Prh
Ent
AIV
CA1
S
CA2
DG
CA3
CA
4
PaS
st
opt
CI
En
CPu
Pir
lo
acp
GP
ic
cp
AO
Acb
sm
ml
TT
SHy
f
mt
fr
Tu
VDB

Fig. 34

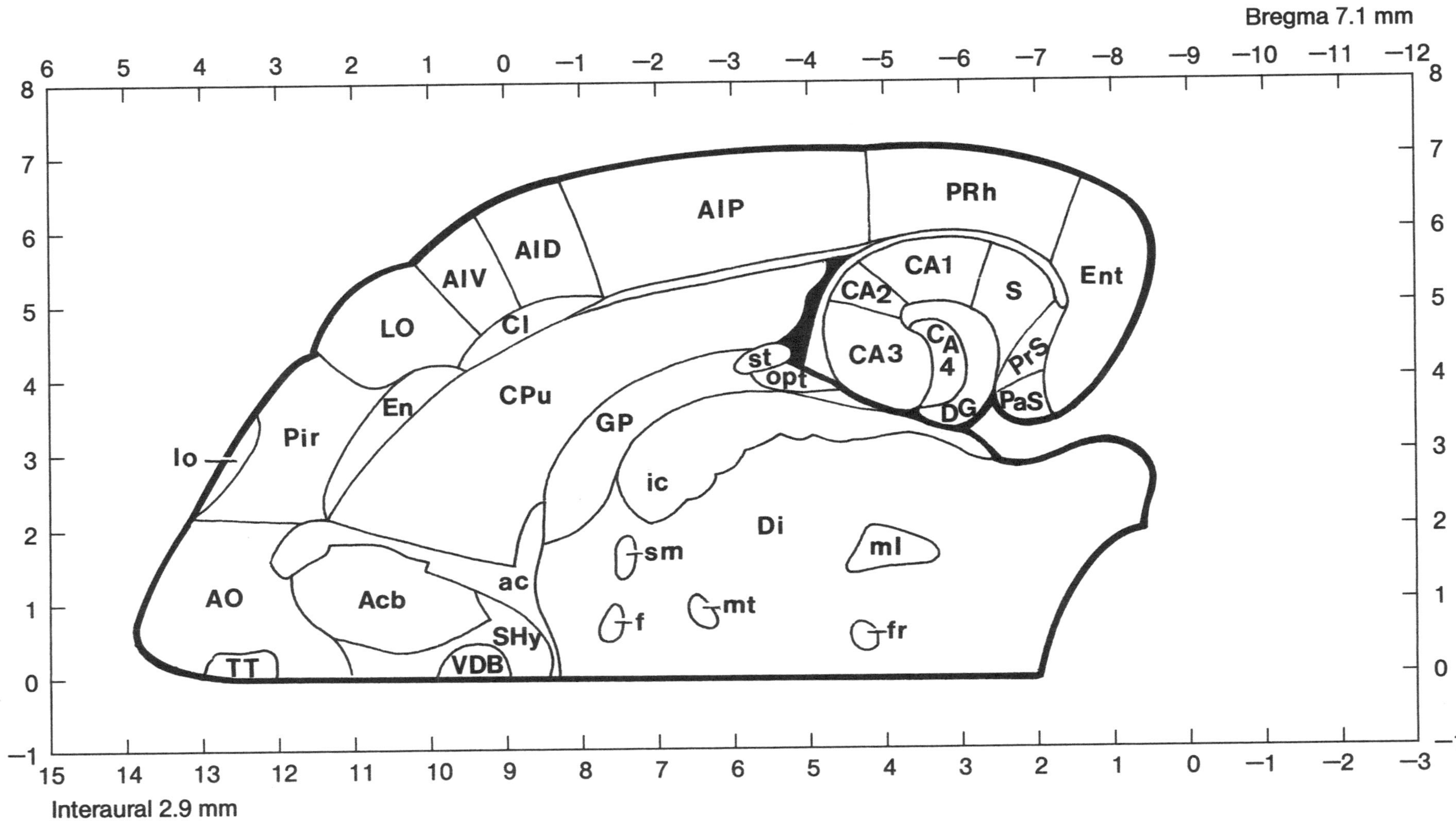

Bregma 7.1 mm
Interaural 2.9 mm
AIP
PRh
AID
AIV
LO
CI
Ent
CA1
CA2
S
CA3
CA
4
PrS
PaS
DG
st
opt
CPu
En
Pir
lo
GP
ic
Di
ml
sm
mt
f
fr
ac
AO
Acb
SHy
VDB
TT

Fig. 35

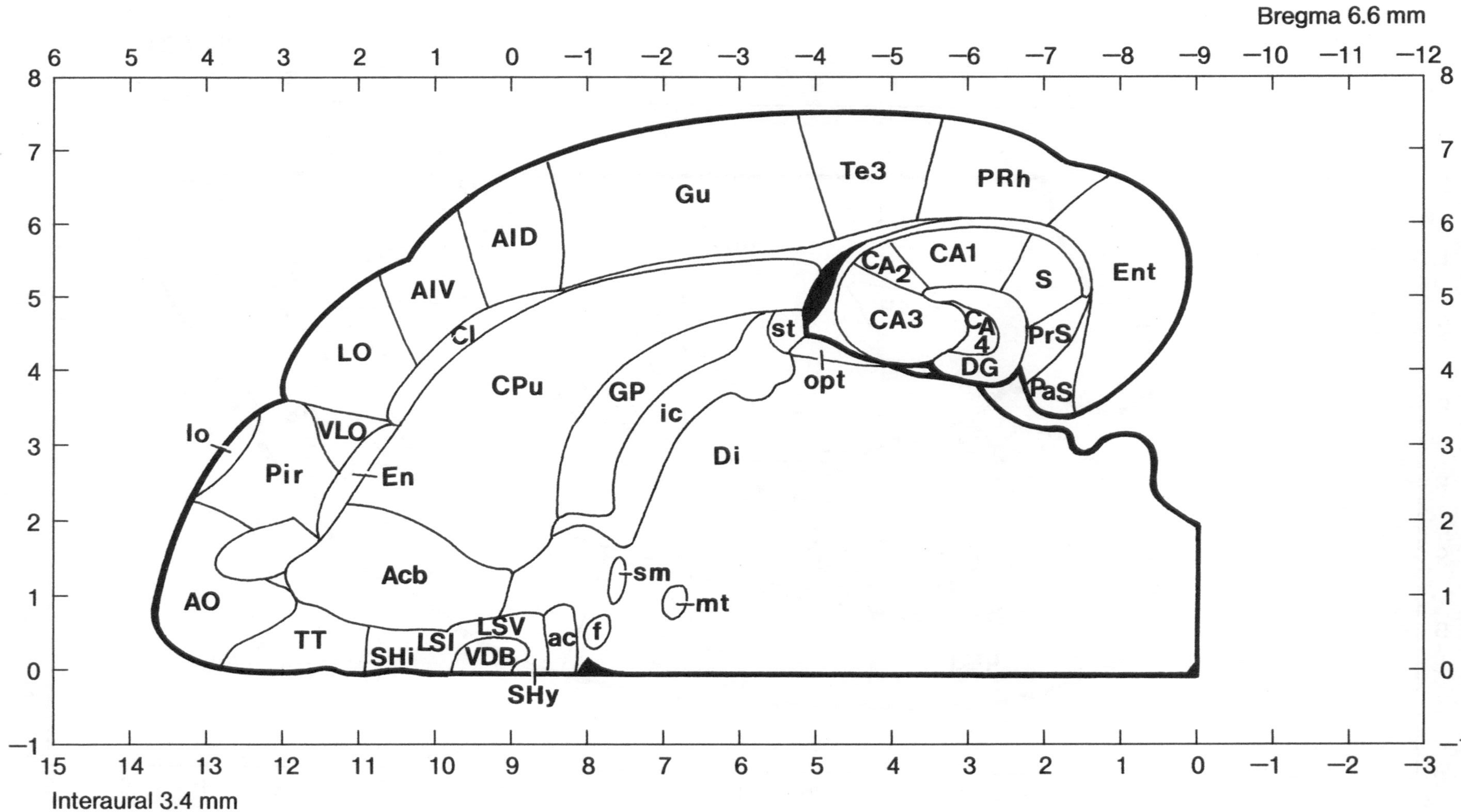
Bregma 6.6 mm
Interaural 3.4 mm
Te3
PRh
Gu
AID
AIV
Ent
CA1
CA2
S
CA3
CA 4
PrS
DG
PaS
st
opt
LO
Cl
CPu
GP
ic
Di
lo
VLO
Pir
En
Acb
sm
mt
AO
TT
SHi
LSI
LSV
VDB
ac
f
SHy

Fig. 36

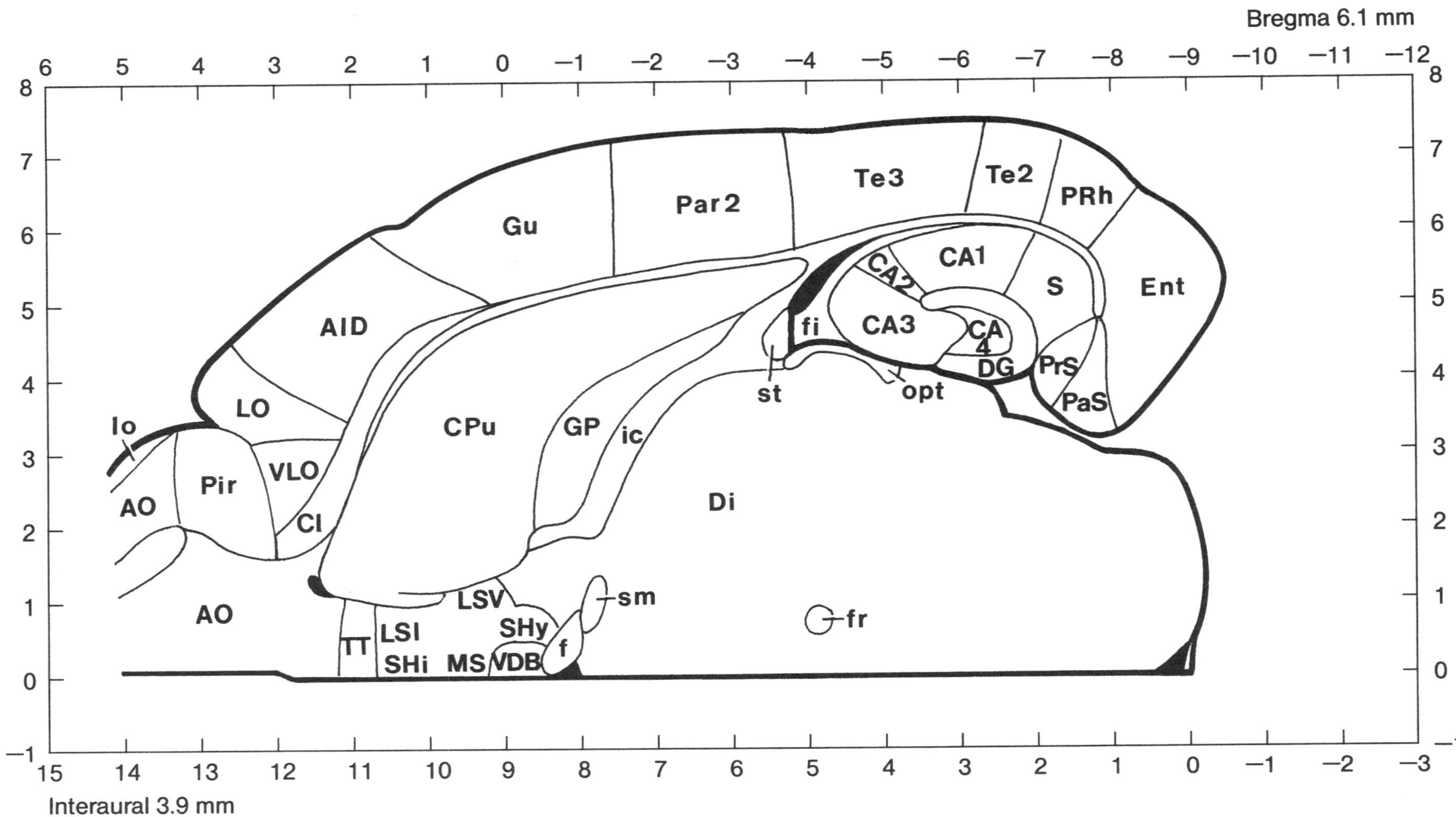
Bregma 6.1 mm
Interaural 3.9 mm
Gu
Par 2
Te3
Te2
PRh
Ent
AID
LO
lo
Pir
VLO
AO
CI
AO
CPu
GP
ic
Di
fi
CA3
CA2
CA1
S
CA 4
DG
PrS
PaS
st
opt
LSV
LSI
TT
SHi
MS
VDB
SHy
f
sm
fr

Fig. 37

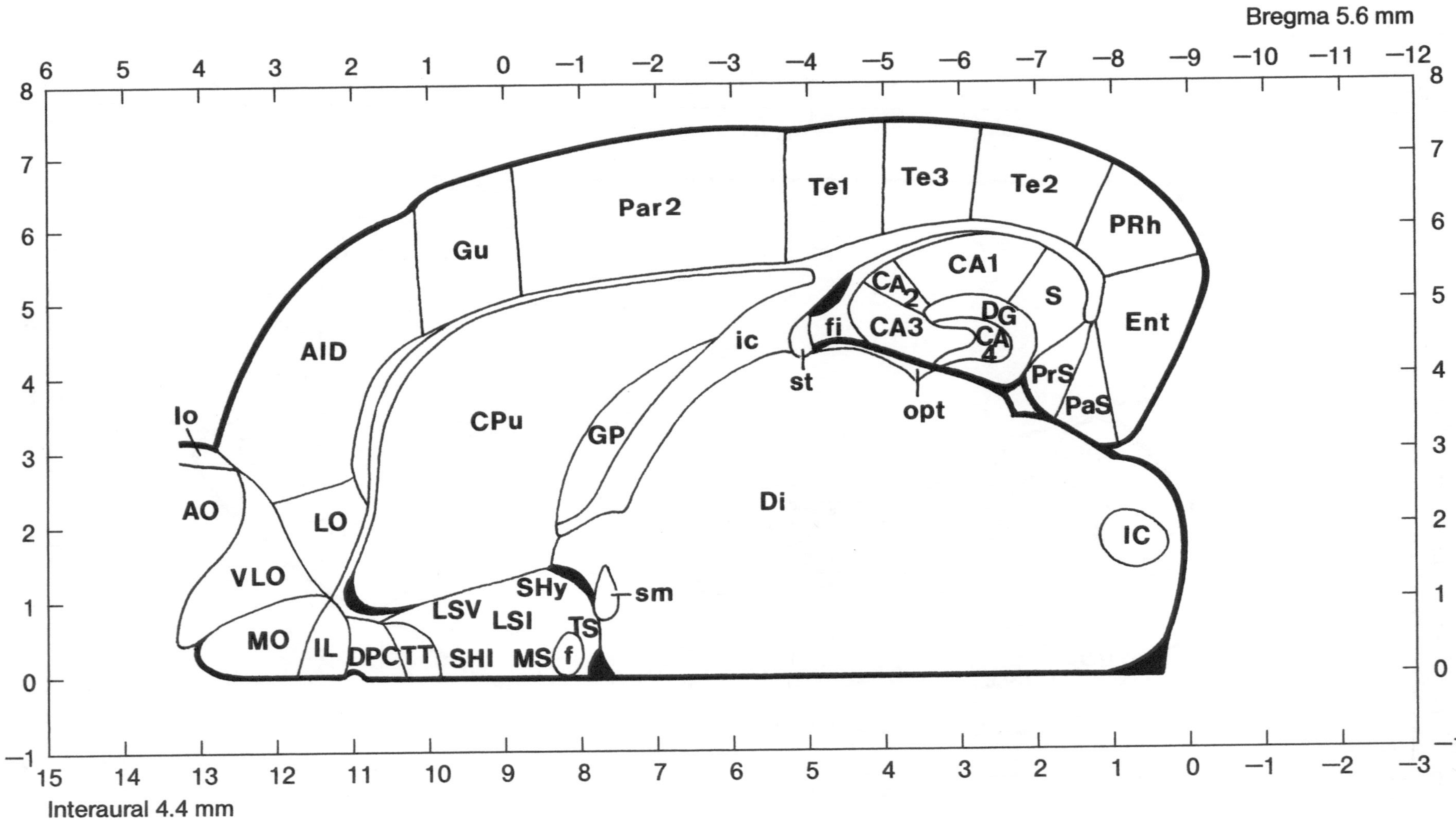
Bregma 5.6 mm
Interaural 4.4 mm
Par2
Te1
Te3
Te2
PRh
Gu
CA1
CA2
S
Ent
DG
CA3
CA4
fi
ic
AID
st
PrS
opt
PaS
lo
CPu
GP
Di
AO
LO
IC
VLO
SHy
sm
LSV
LSI
TS
MO
IL
DPC
TT
SHI
MS
f

Fig. 38

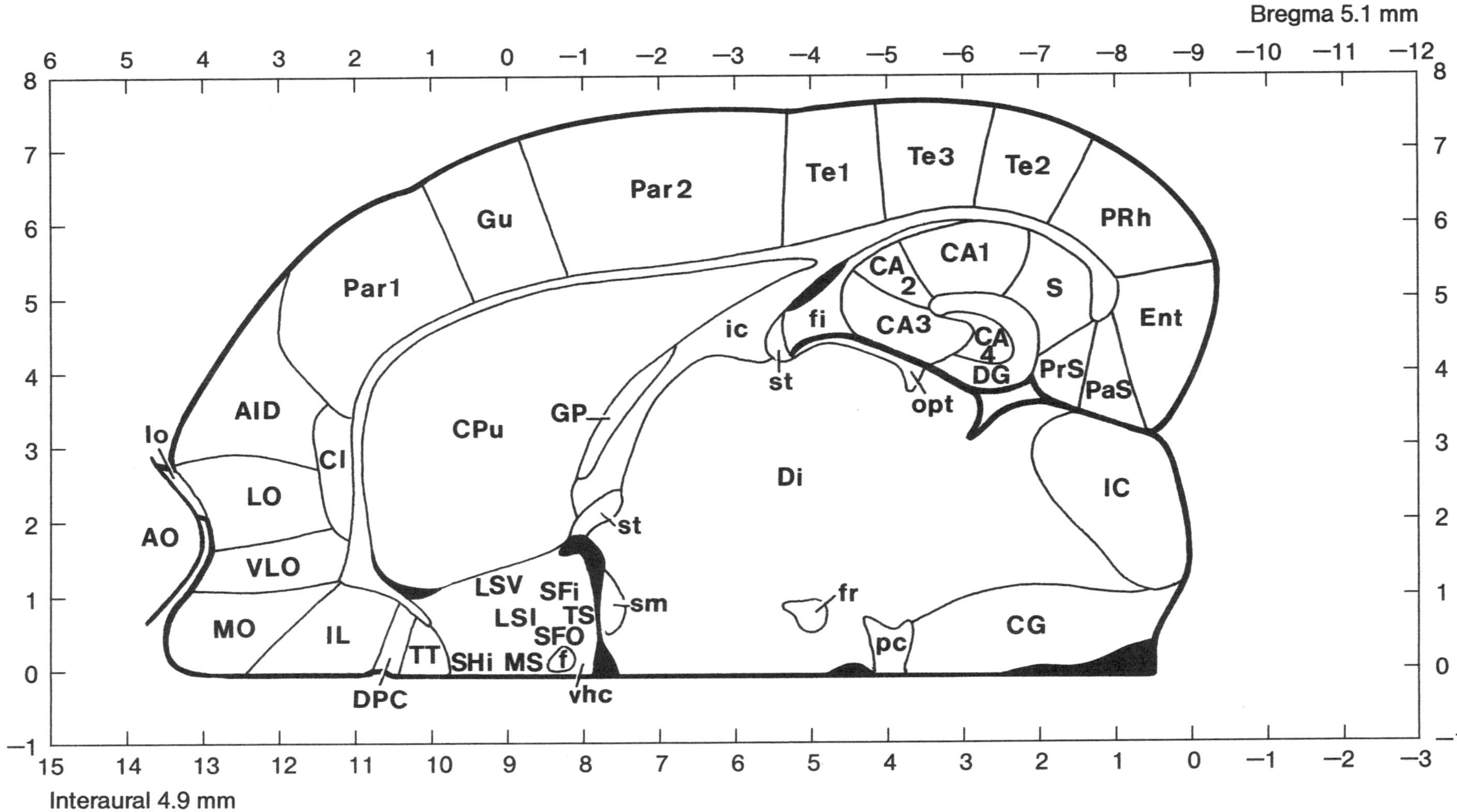
Bregma 5.1 mm
Interaural 4.9 mm
PRh
Ent
PaS
PrS
Te2
Te3
Te1
S
CA1
CA 2
CA3
CA 4
DG
opt
fi
st
ic
Di
IC
CG
pc
fr
Par2
Gu
Par1
CPu
GP
st
sm
SFi
TS
SFO
f
vhc
LSV
LSI
MS
SHi
TT
DPC
IL
CI
AID
LO
VLO
MO
AO
lo

Fig. 39

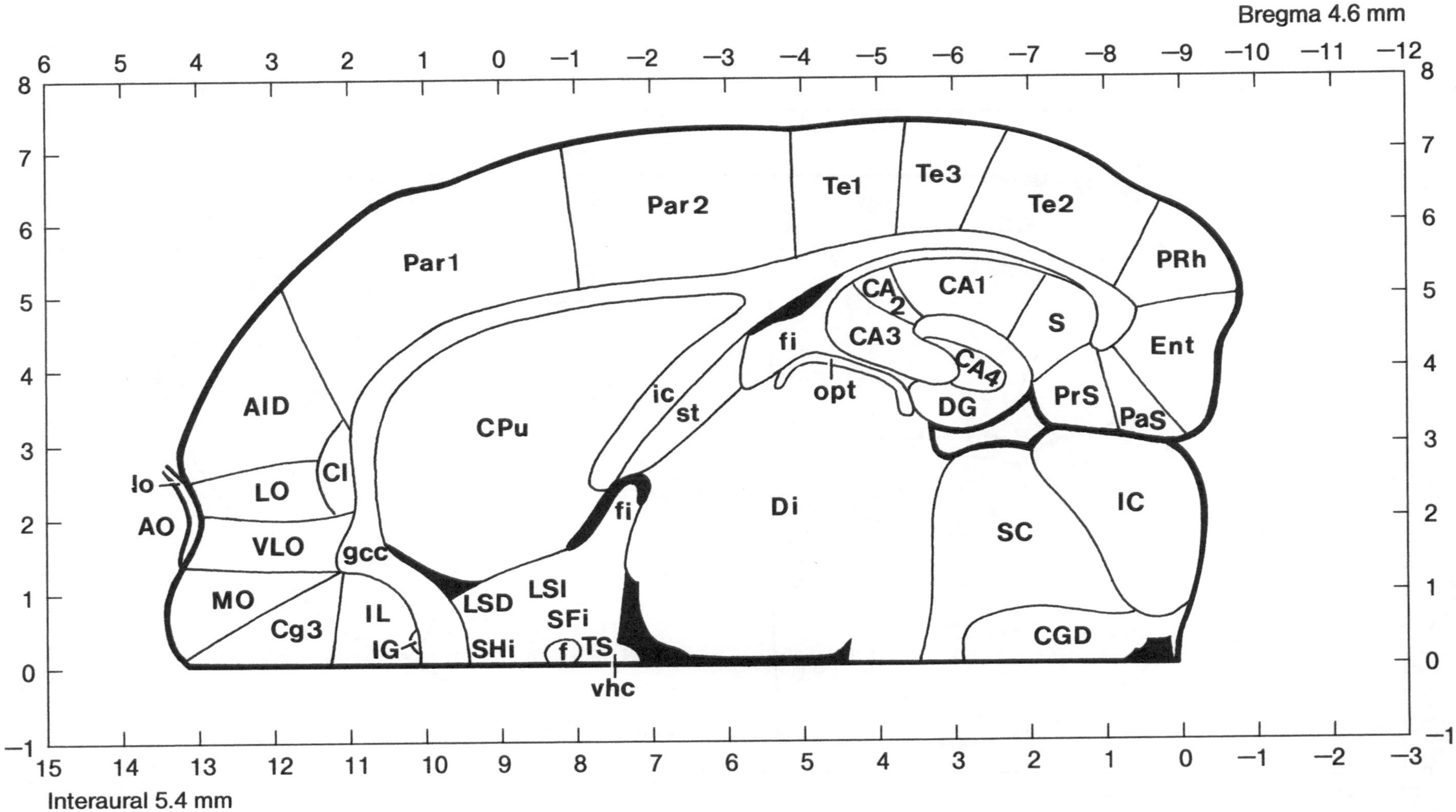
Bregma 4.6 mm
Par1
Par2
Te1
Te3
Te2
PRh
Ent
PaS
PrS
S
CA1
CA2
CA3
CA4
DG
fi
opt
ic
st
CPu
AID
LO
Cl
lo
AO
VLO
gcc
MO
Cg3
IL
IG
LSD
LSI
SFi
SHi
f
TS
vhc
Di
SC
IC
CGD
Interaural 5.4 mm

Fig. 40

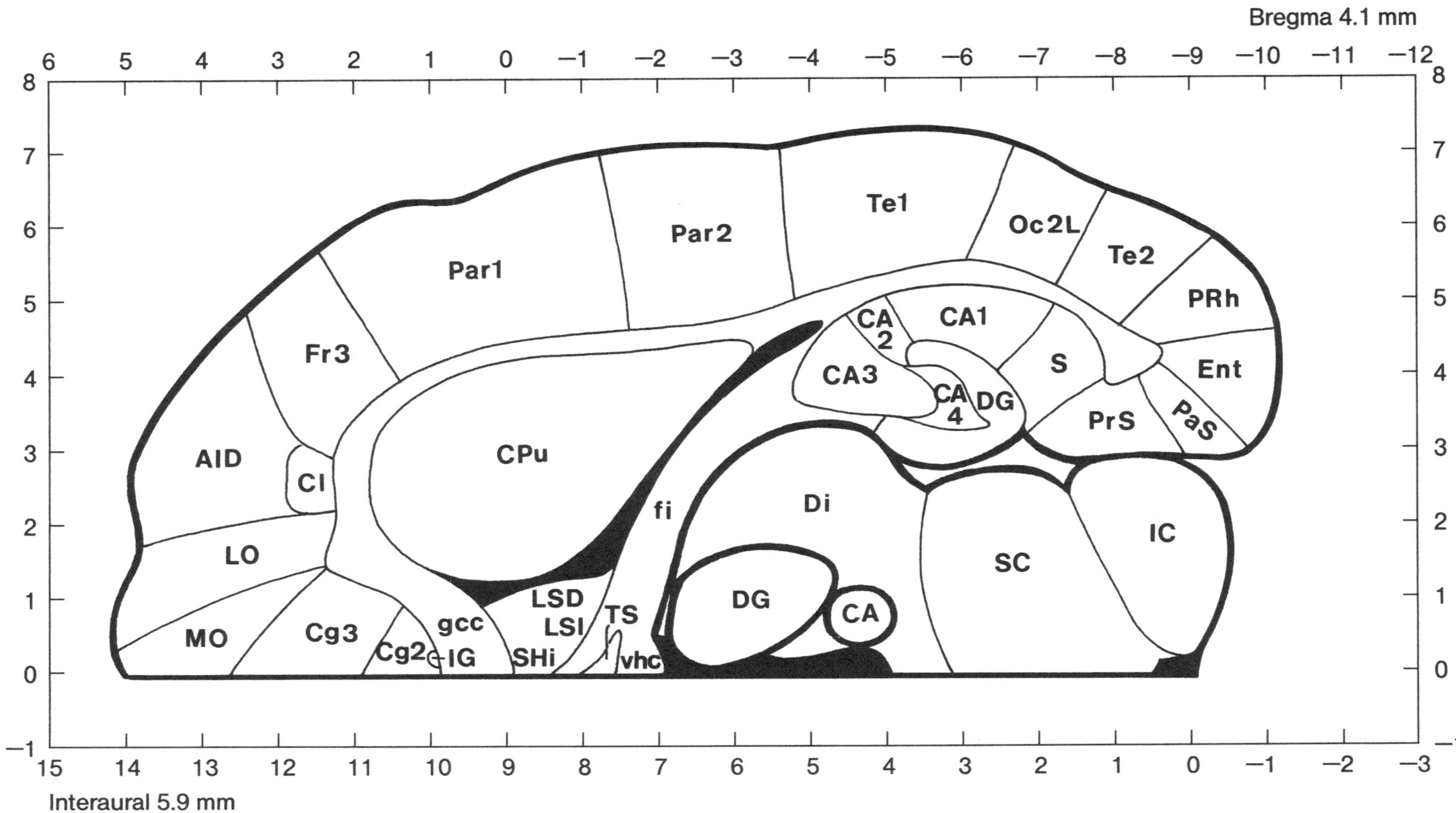
Bregma 4.1 mm
Interaural 5.9 mm
Par1
Par2
Te1
Oc2L
Te2
PRh
Ent
PaS
PrS
S
CA1
CA 2
CA3
CA 4
DG
Fr3
AID
CI
LO
MO
Cg3
Cg2
IG
gcc
CPu
fi
LSD
LSI
SHi
TS
vhc
Di
DG
CA
SC
IC

Fig. 41

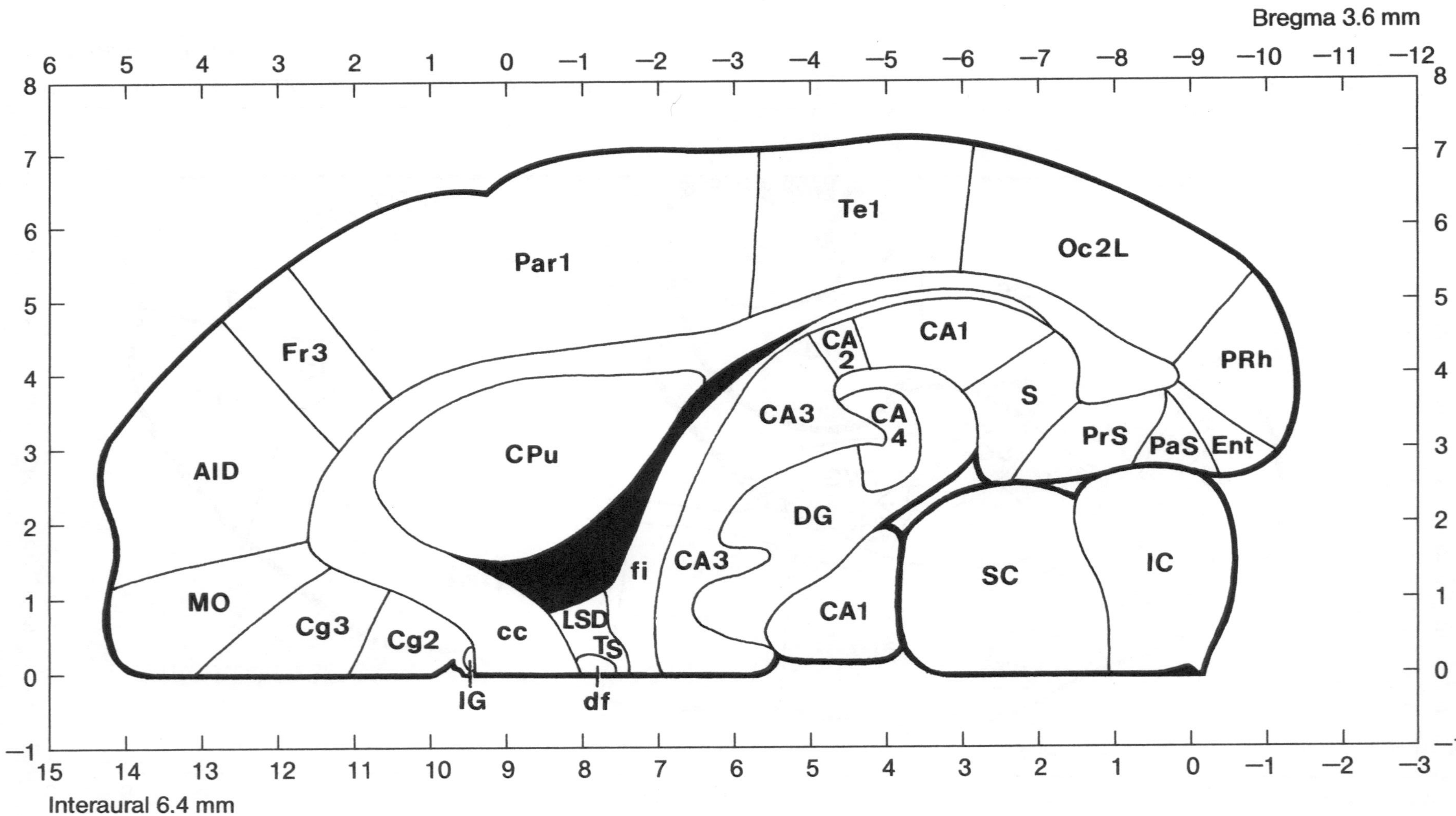
Bregma 3.6 mm
Interaural 6.4 mm
Par1
Te1
Oc2L
PRh
Fr3
AID
MO
Cg3
Cg2
cc
IG
df
LSD
TS
fi
CPu
CA3
CA
2
CA1
CA
4
DG
CA3
CA1
S
PrS
PaS
Ent
SC
IC
6
5
4
3
2
1
0
−1
−2
−3
−4
−5
−6
−7
−8
−9
−10
−11
−12
15
14
13
12
11
10
9
8
7
6
5
4
3
2
1
0
−1
−2
−3
8
7
6
5
4
3
2
1
0
−1

Fig. 42

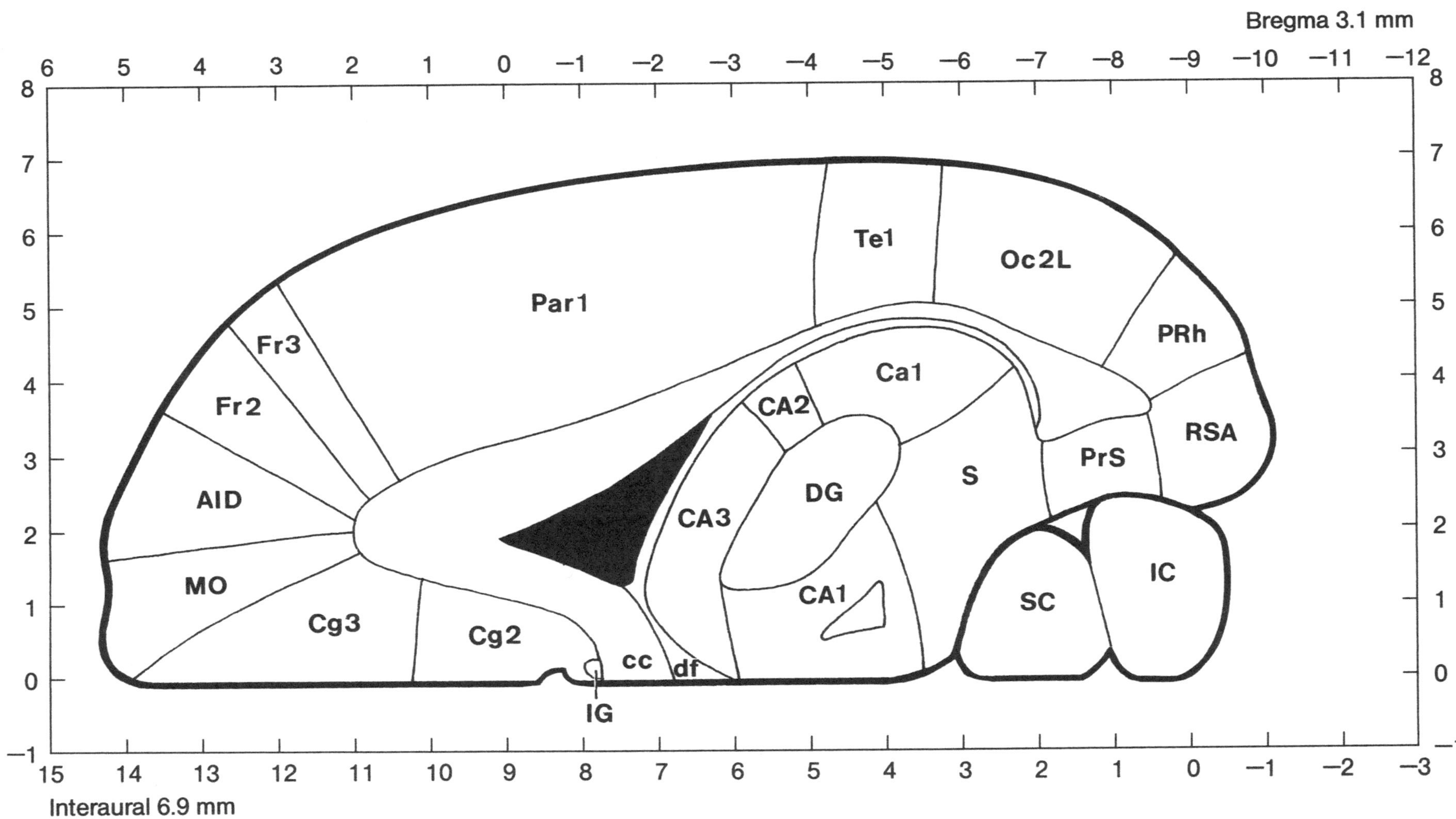
Bregma 3.1 mm
6 5 4 3 2 1 0 −1 −2 −3 −4 −5 −6 −7 −8 −9 −10 −11 −12
8 7 6 5 4 3 2 1 0 −1
Par1
Te1
Oc2L
PRh
RSA
Fr3
Fr2
AID
MO
Cg3
Cg2
CA2
Ca1
DG
S
PrS
CA3
CA1
SC
IC
cc
df
IG
15 14 13 12 11 10 9 8 7 6 5 4 3 2 1 0 −1 −2 −3
Interaural 6.9 mm

Fig. 43

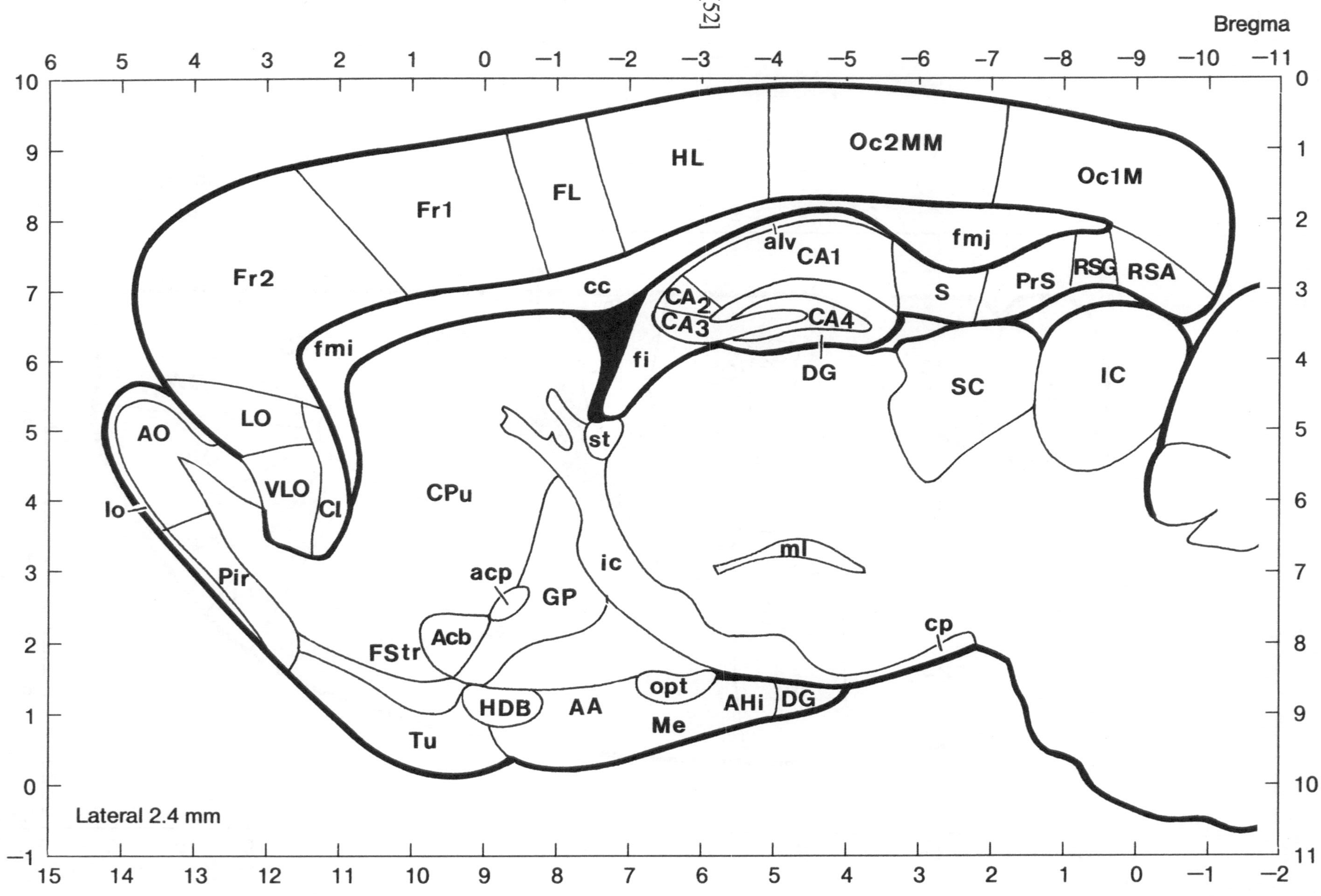
Bregma
6
5
4
3
2
1
0
−1
−2
−3
−4
−5
−6
−7
−8
−9
−10
−11
Oc2MM
Oc1M
HL
FL
Fr1
Fr2
fmj
alv
CA1
RSG
RSA
PrS
S
cc
CA2
CA3
CA4
fmi
fi
DG
SC
IC
LO
AO
st
VLO
Cl
CPu
lo
ml
ic
acp
Pir
GP
Acb
cp
FStr
opt
HDB
AA
AHi
DG
Me
Tu
Lateral 2.4 mm
Interaural
15
14
13
12
11
10
9
8
7
6
5
4
3
2
1
0
−1
−2

Fig. 44

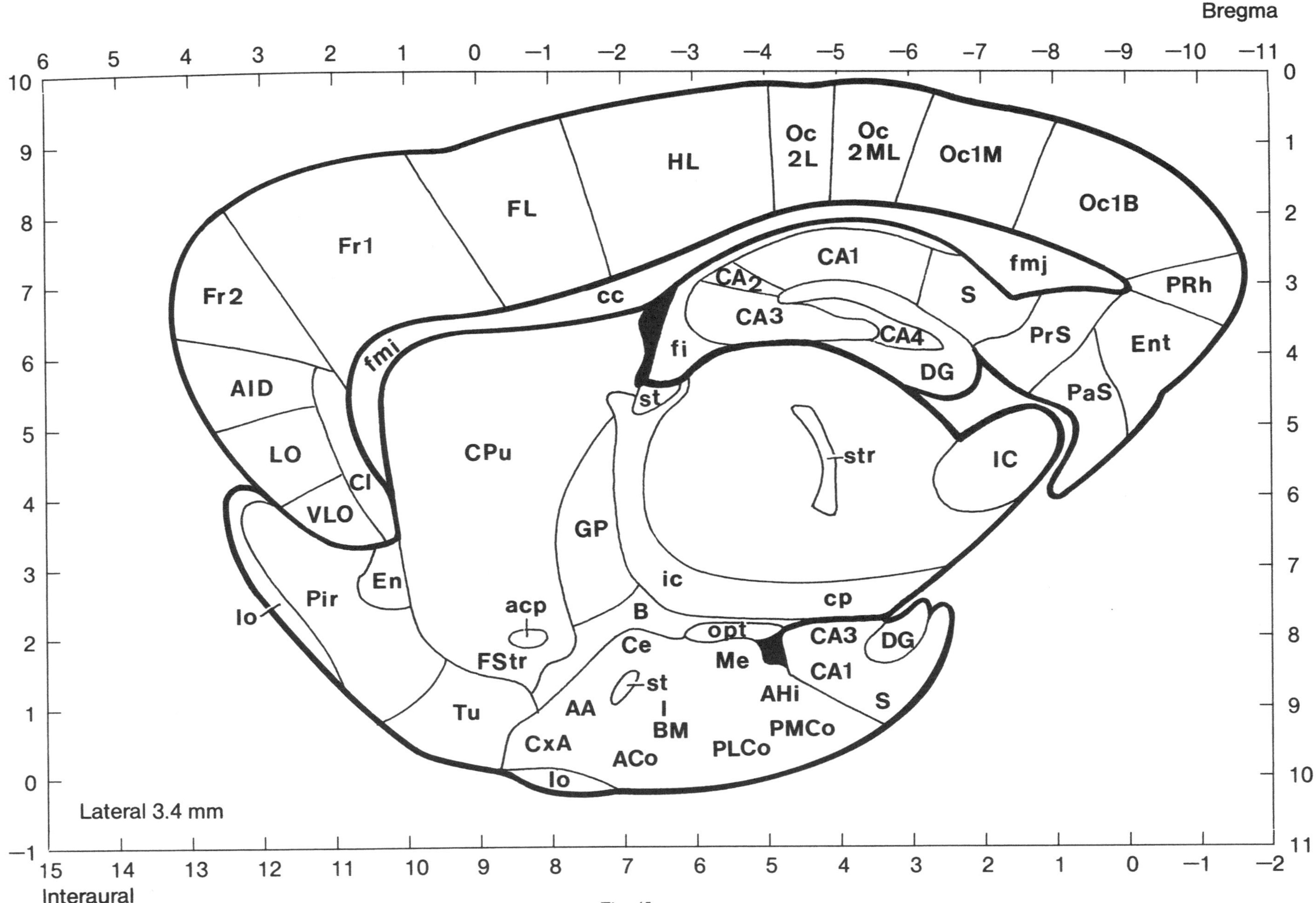
Bregma
Interaural
Lateral 3.4 mm
Oc1B
Oc1M
Oc 2ML
Oc 2L
HL
FL
Fr1
Fr2
AID
LO
VLO
Cl
fmi
cc
CPu
En
Pir
lo
acp
FStr
Tu
GP
ic
cp
B
Ce
st
AA
CxA
ACo
BM
I
Me
opt
AHi
PLCo
PMCo
CA3
CA1
DG
S
str
IC
fi
CA2
CA1
CA3
CA4
DG
S
fmj
PrS
PaS
Ent
PRh

Fig. 45

Cortical Maps in Stereotaxic Coordinates

The cortical maps in Figs. 46–49 have been prepared from the coronal (Figs. 2–31) and horizontal (Figs. 32–43) sections by orthogonal projection. This provides lateral (Figs. 46 and 49), dorsal (Fig. 47), and medial (Fig. 48) views of the rat cortex in stereotaxic coordinates. Figure 49 shows only part of the lateral view, because the most dorsal and basal sections of the horizontal series do not admit of complete reconstruction because of the tangential sections at the top and bottom of the hemisphere. In these tangential sections a reliable delineation of cortical areas is not possible. All these cortical maps show the distances in millimeters from bregma or the interaural line at the top or bottom margins. The distance from the horizontal plane passing through the interaural line and the distance from the horizontal plane passing through bregma and lambda on the surface of the skull are indicated in the left and right margins, respectively, for both the lateral (Figs. 46 and 49) and the medial (Fig. 48) views. The distance in millimeters from the midline between both hemispheres is given for the dorsal view (Fig. 47) in the left and right margins.

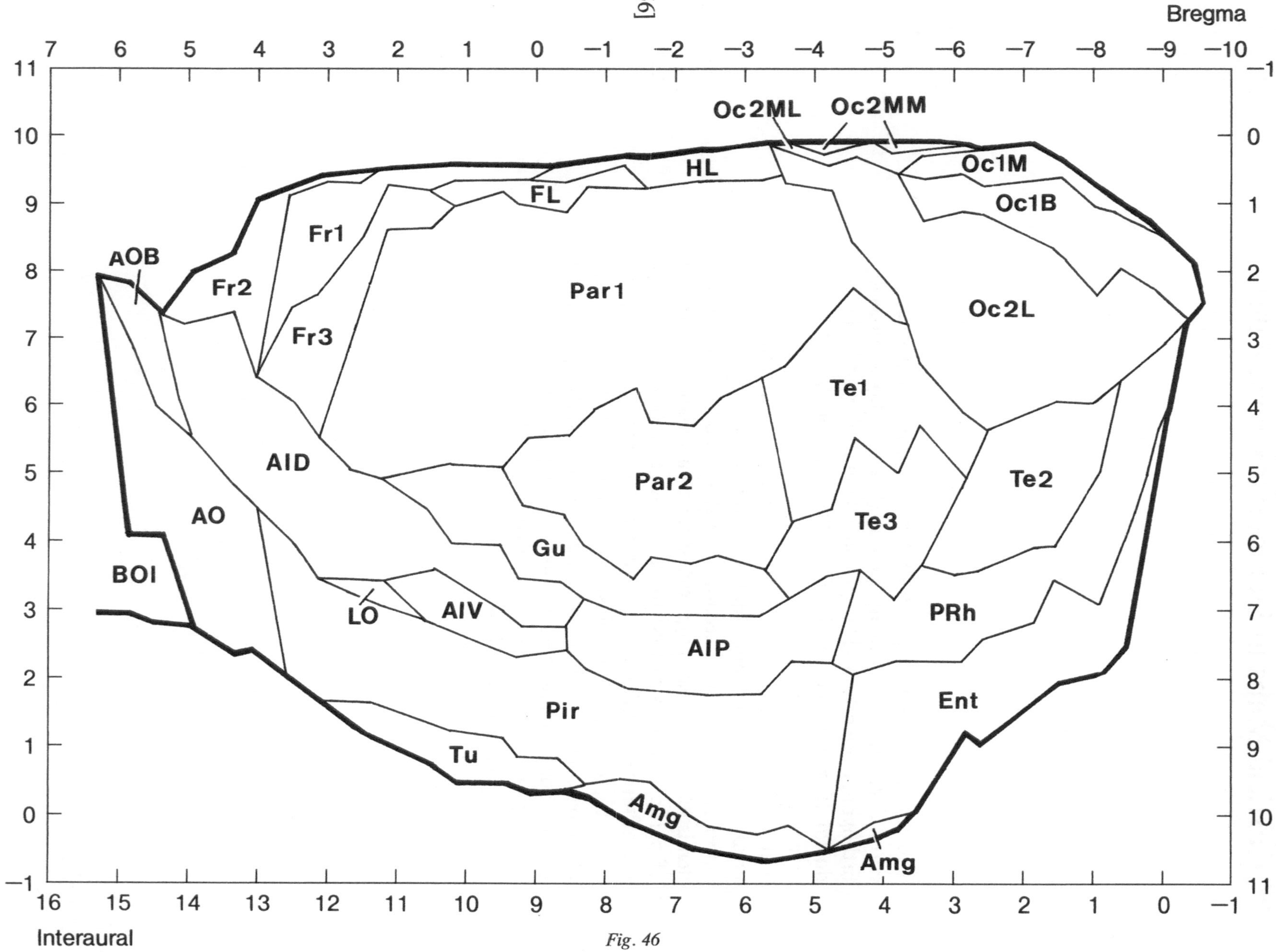

Bregma
Interaural
AOB
Fr2
Fr1
Fr3
Par1
FL
HL
Oc2ML
Oc2MM
Oc1M
Oc1B
Oc2L
Te1
Te2
Te3
Par2
AID
AO
BOI
Gu
LO
AIV
AIP
PRh
Ent
Pir
Tu
Amg
Amg

Fig. 46

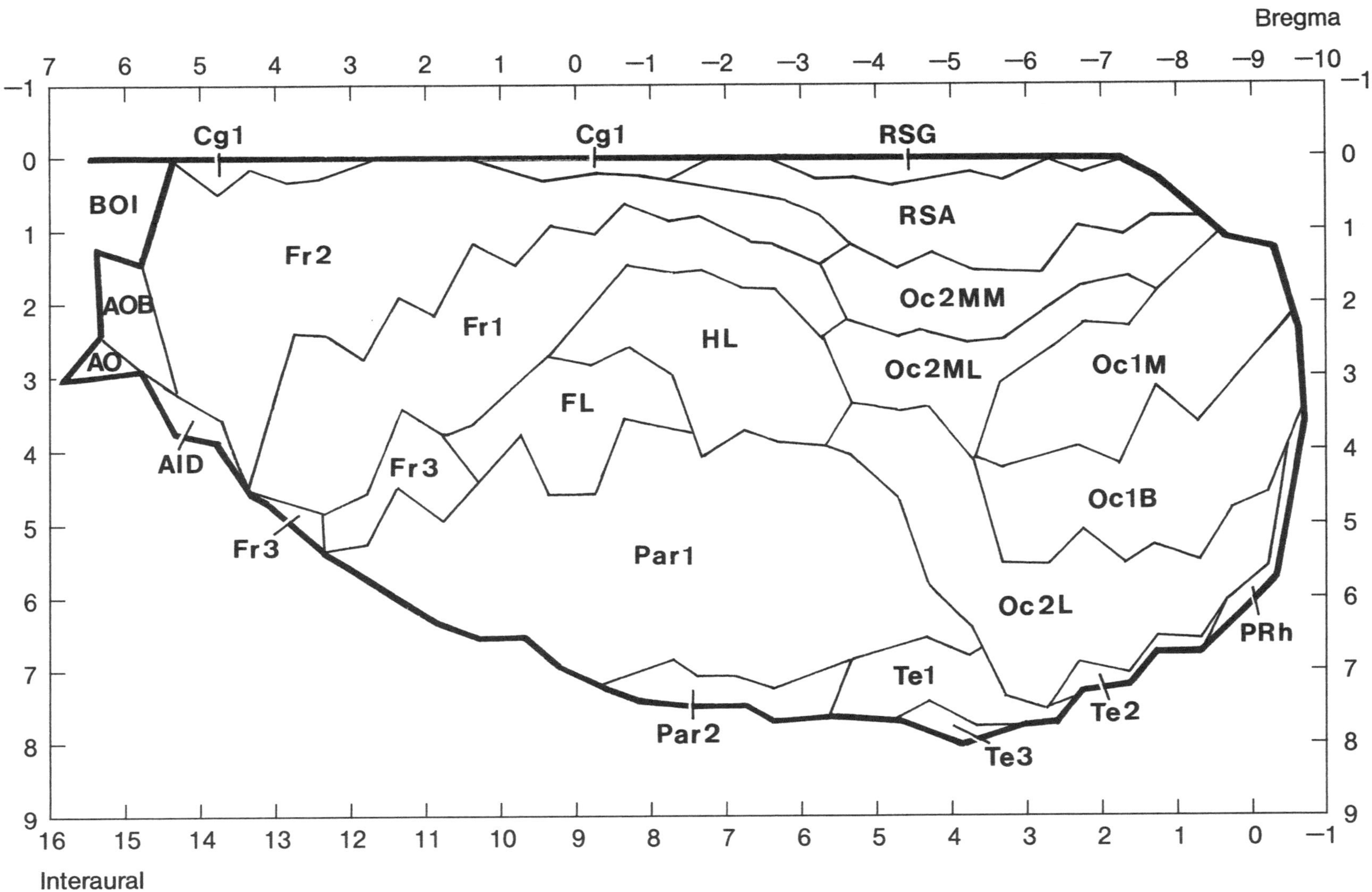
Bregma
Interaural
Cg1
Cg1
RSG
BOI
AOB
AO
AID
Fr3
Fr2
Fr1
Fr3
FL
HL
RSA
Oc2MM
Oc2ML
Oc1M
Oc1B
Oc2L
Par1
Par2
Te1
Te2
Te3
PRh

Fig. 47

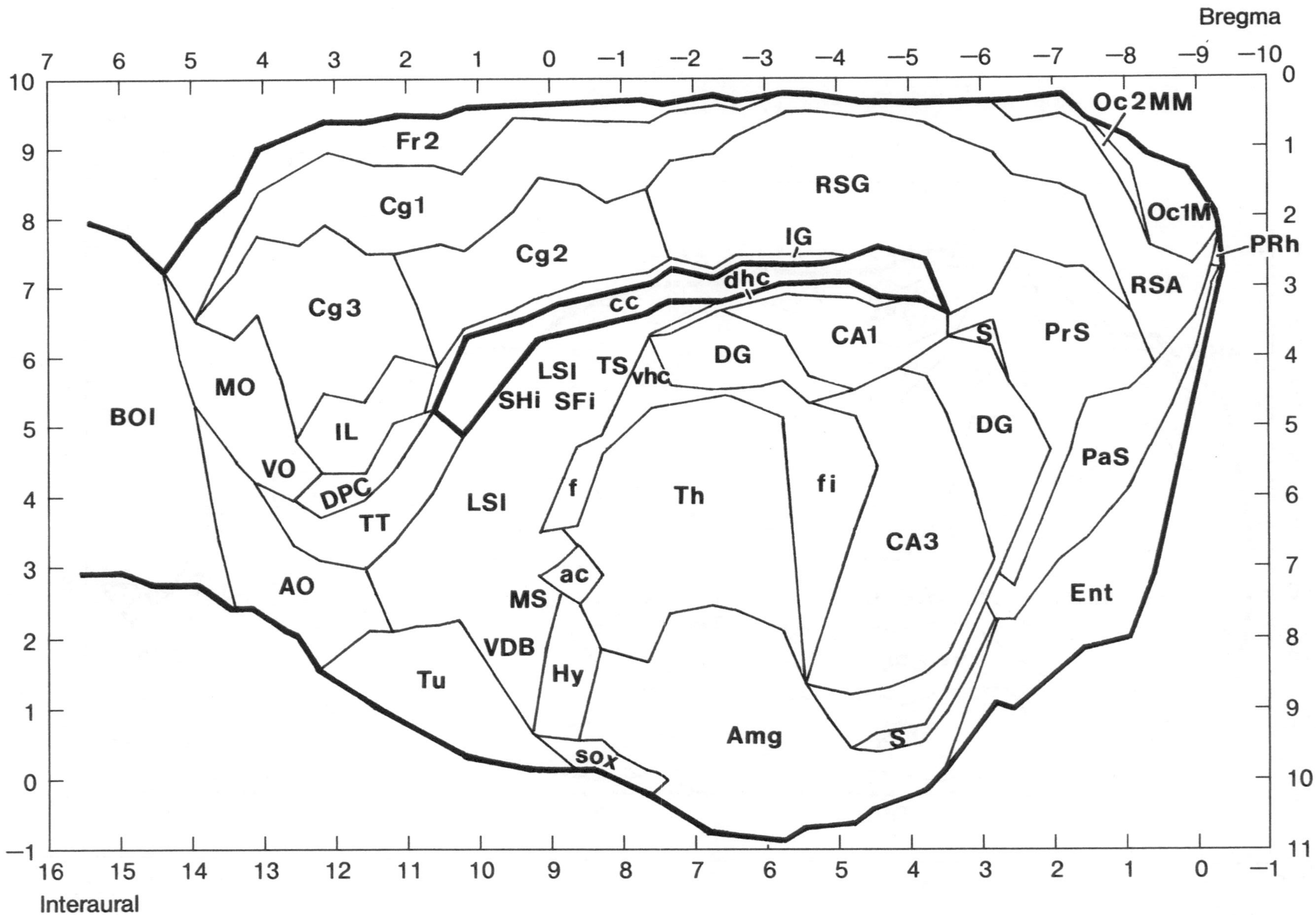
Bregma
Interaural
Oc2MM
Oc1M
PRh
RSA
RSG
Fr2
Cg1
Cg2
Cg3
IG
dhc
cc
CA1
S
PrS
DG
LSI
TS
vhc
SHi
SFi
MO
BOI
IL
VO
DPC
TT
f
Th
fi
CA3
PaS
Ent
ac
MS
AO
VDB
Hy
Tu
Amg
sox

Fig. 48

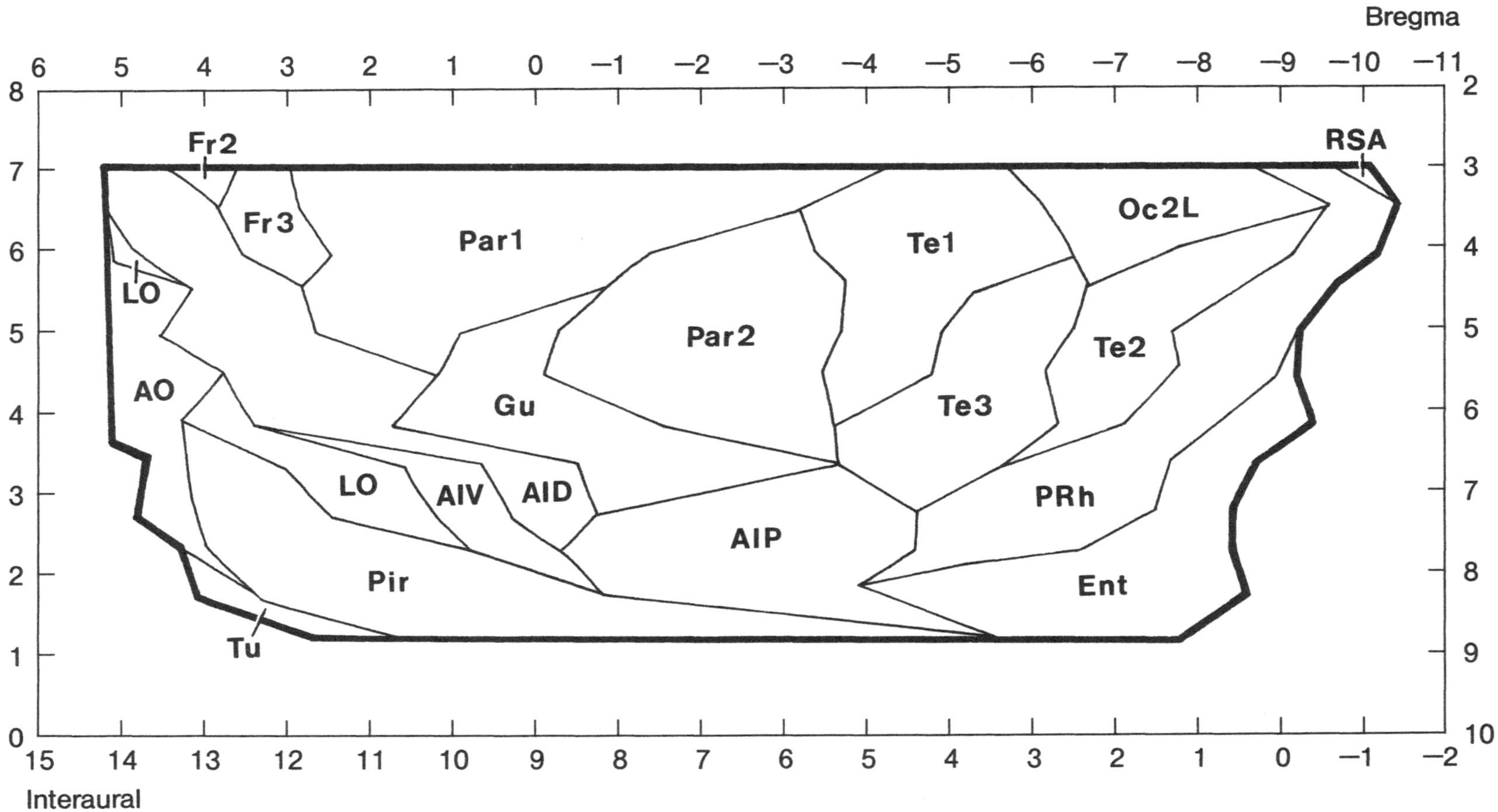
Bregma
6
5
4
3
2
1
0
−1
−2
−3
−4
−5
−6
−7
−8
−9
−10
−11
8
7
6
5
4
3
2
1
0
2
3
4
5
6
7
8
9
10
15
14
13
12
11
10
9
8
7
6
5
4
3
2
1
0
−1
−2
Interaural
Fr2
RSA
Fr3
Par1
Te1
Oc2L
LO
Par2
Te2
AO
Gu
Te3
LO
AIV
AID
PRh
AIP
Pir
Ent
Tu

Fig. 49

Frontal and Caudal Aspects of the Hemisphere in Computer Reconstructions

The most frontal and caudal parts of the hemispheres can be only poorly demonstrated in orthogonal cortical maps. The coronal drawings (Figs. 2–31) were therefore used for computer reconstruction with a 3-D program in views at varying angles (Figs. 50–54). The frontal and caudal poles are shown in Figs. 50 and 52, respectively, with no rotation, solely in the frontal and caudal views. In both cases the stereotaxic coordinates are given for the first and last sections as distances from bregma in millimeters. In addition, the distances of structures from the midline are shown at the top and bottom. The numbers in the right [61]
margin show the dorsoventral distance from the horizontal plane passing through bregma and lambda on the surface of the skull.

Figs. 51 and 53 show aspects of the frontal and caudal poles of the hemisphere from a medial and dorsal standpoint. The hemisphere has been rotated clockwise through 80° around a sagittal axis, and this axis has also been tilted through 50° in the horizontal plane.

Figure 54 shows the caudal pole of the hemisphere in a lateral and dorsal view. In this case the hemisphere has been rotated counterclockwise through 60° around the sagittal axis and tilted through 50° in the horizontal plane. The areal borders in these 3-D reconstructions are drawn with thick lines, and the sectioning plane of the first section in rostral or caudal views is crosshatched.

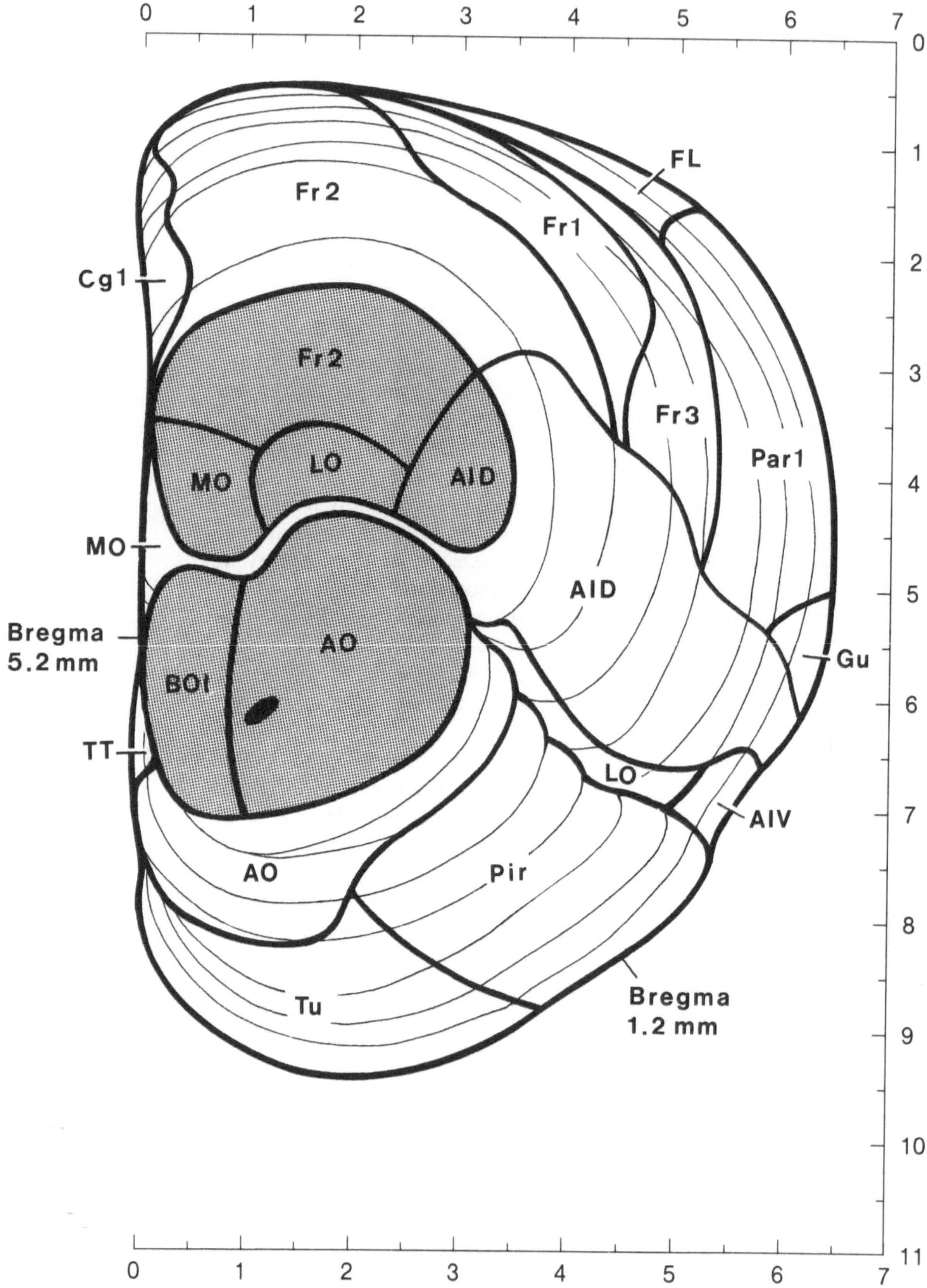

FL
Fr2
Fr1
Cg1
Fr2
Fr3
Par1
MO
LO
AID
MO
AID
Bregma
5.2 mm
AO
Gu
BOI
TT
LO
AIV
AO
Pir
Tu
Bregma
1.2 mm

[62]

Fig. 50

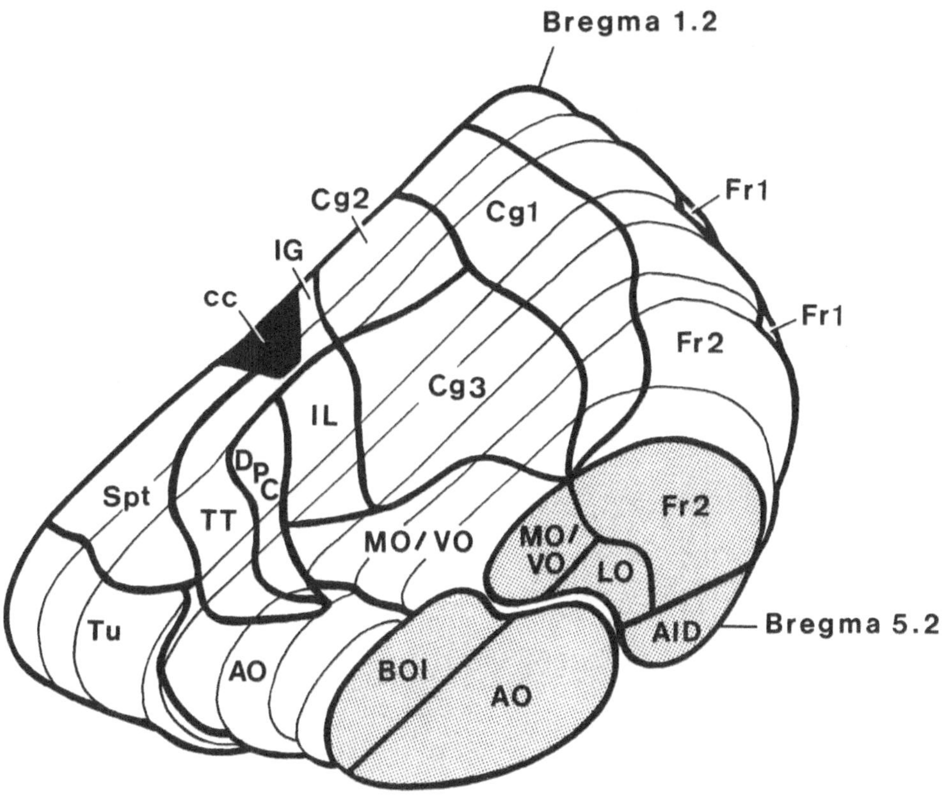

Bregma 1.2
Cg2
Cg1
Fr1
IG
cc
Fr1
Fr2
Cg3
IL
D
P
C
Spt
TT
Fr2
MO/VO
MO/
VO
LO
Tu
AO
BOI
AO
AID
Bregma 5.2

[63]

Fig. 51

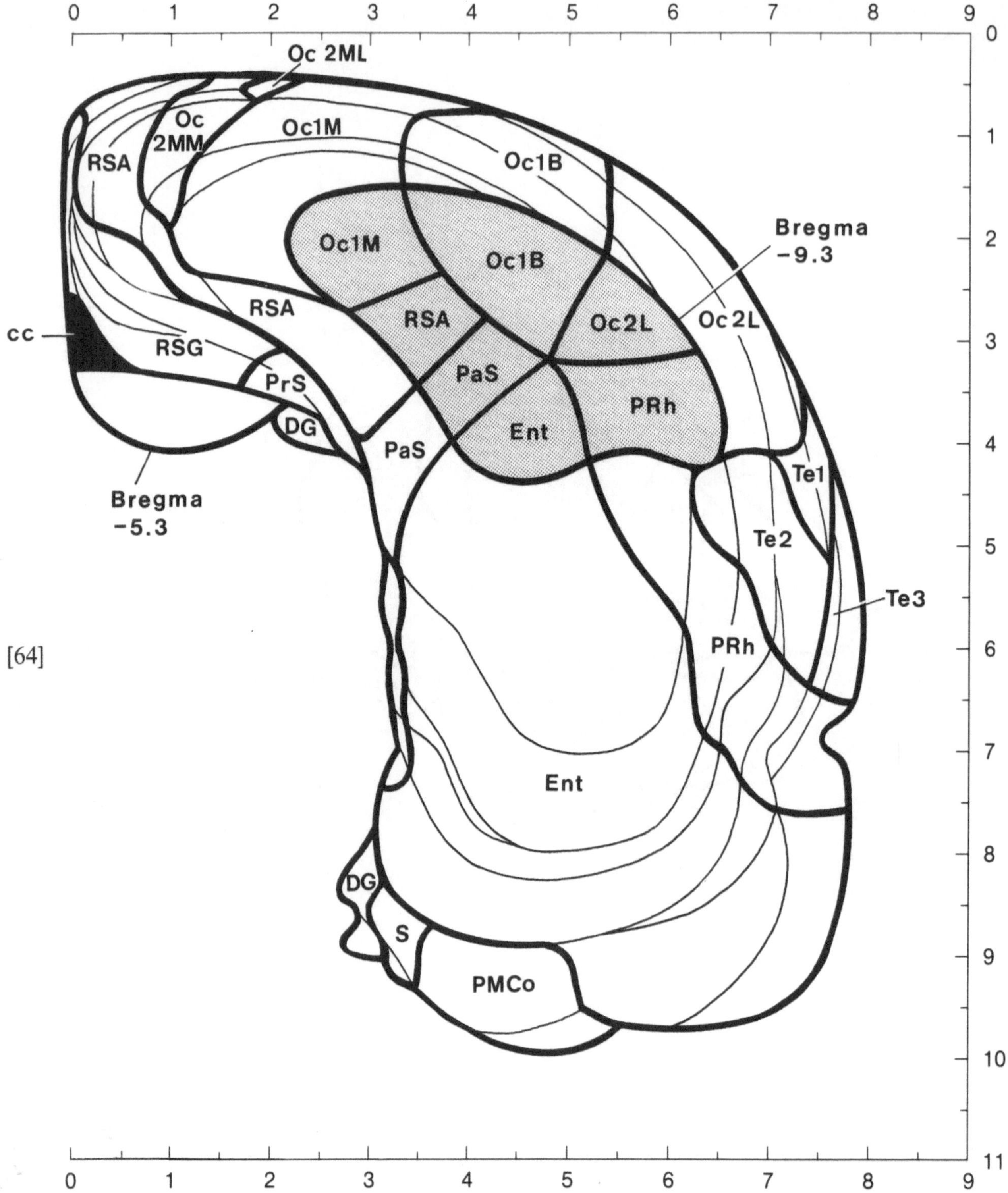

Oc 2ML
Oc
2MM
Oc1M
RSA
Oc1B
Oc1M
Oc1B
Bregma
−9.3
RSA
RSA
Oc2L
Oc2L
cc
RSG
PaS
PrS
PRh
DG
Ent
PaS
Te1
Bregma
−5.3
Te2
Te3
PRh
Ent
DG
S
PMCo

[64]

Fig. 52

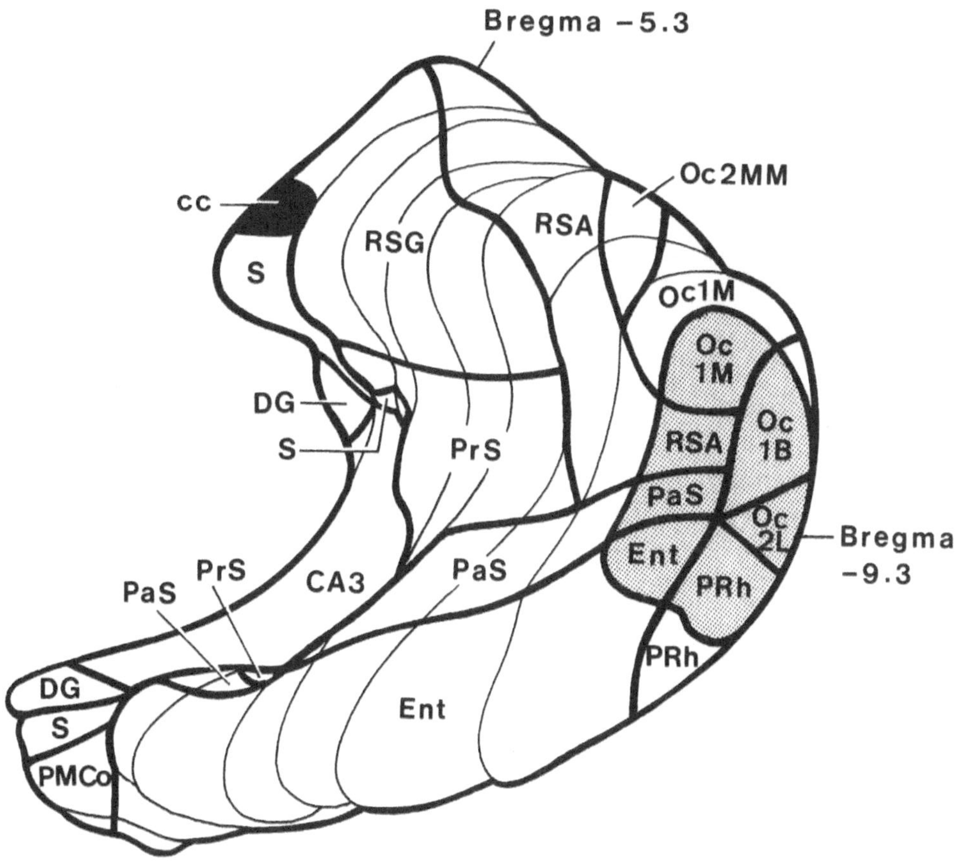
Bregma −5.3
Oc2MM
cc
RSA
RSG
S
Oc1M
Oc
1M
DG
Oc
1B
RSA
S
PrS
PaS
Oc
2L
Bregma
−9.3
Ent
PaS
PrS
PaS
CA3
PRh
PRh
DG
Ent
S
PMCo

[65]

Fig. 53

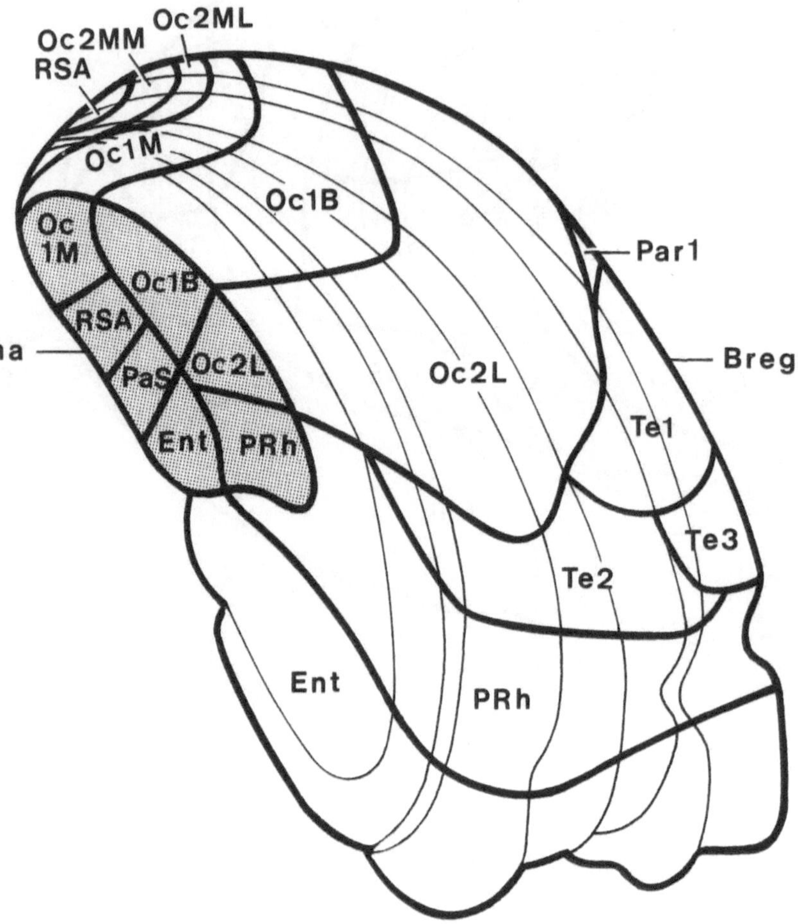
Oc2MM
Oc2ML
RSA
Oc1M
Oc1B
Oc
1M
Oc1B
RSA
PaS
Oc2L
Ent
PRh
Bregma
-9.3
Par1
Bregma -5.3
Oc2L
Te1
Te3
Te2
Ent
PRh

[66]

Fig. 54

Delineation of Cortical Areas in Nissl- and Myelin-Stained Sections

The rat brains from which the sections presented in the micrographs in this section (Figs. 55–67) were taken were processed differently than the brains in the Paxinos and Watson (1982) atlas. Most of the brains used in the present atlas came from animals placed in the stereotaxic apparatus after being perfused with fixative.

After perfusion of the animals (cf. p. 12) the skulls were partially removed, but the entire bases of the skulls were left completely intact. Following these preparations, the specimens were placed in a stereotaxic apparatus in the flat-skull position. [67]
The brains were then sectioned in a coronal plane with a razor blade fixed to an electrode holder positioned at the interaural line. The two fragments of each brain were embedded in paraffin and serial 20-μm sections were cut. These sections were stained alternately with Nissl and myelin. The planes in which the sections lie can be deduced from a comparison with the plates in the Paxinos and Watson (1982) atlas, using the bregma and interaural reference points for orientation. This procedure provides sections of good histological quality with known stereotaxic positions. A comparison of the plates in the Paxinos and Watson (1982) atlas with the micrographs of the present material reveals striking compatibility, and in some cases almost complete correspondence. The coronal sections of Paxinos and Watson's specimens show only two distinct deviations from these specimens in Figs. 55–67. The coronally sectioned brain in the Paxinos and Watson (1982) atlas is somewhat broader in the area of the occipital cortex than the Wistar rat brains depicted in Figs. 55–67. This variation may be due to the different stock from which the Wistar strains were bred.

The second variation was in the position of the hippocampus along the rostrocaudal axis. In the brain shown in the Paxinos and Watson (1982) atlas the hippocampus appears to be about 0.5 mm further rostrad than in any other Wistar, Lewis, or hooded rat brains processed specifically for the present atlas. Apart from these two differences, all cortical and subcortical features have been found to be at comparable positions along the rostrocaudal axis.

Fixation, Embedding, and Sectioning

While in deep anesthesia the rats were perfused with 50–100 ml Ringer's solution through the left ventricle of the heart, followed by Bodian's fixative (900 ml 80% ethanol, 50 ml 37% formaldehyde, and 50 ml glacial acetic acid at 4° C). About 1 liter of this fixative was used per adult rat over 60 min. The brains were removed 1–3 h after perfusion was complete. They were then stored for 1–2 days in the same fixative.

The brains were dehydrated in a graded ethanol/methylbenzoate series and embedded in paraffin. Once the brains had been carefully blocked, serial 20-µm sections were cut. Gelatin-coated slides were used to mount alternate sections in Nissl and myelin series.

Nissl Staining

The slides were treated in the following baths: xylene, 100% ethanol, 96% ethanol, 70% ethanol, distilled water. They were then stained for 30 min in cresyl fast violet solution at 60° C. To make this solution, a buffer consisting of 100 ml distilled water, 0.544 g sodium acetate, and 0.1 ml 99%–100% acetic acid is added to 2 mg cresyl fast violet. The slides were differentiated by the following procedure: rinsing in distilled water and 70% ethanol (rapid contrast increase) or 96% ethanol (slower contrast increase). Following immersion in xylene the slides were covered with Depex.

Myelin Staining

The slides were prepared for staining in xylene, 100% ethanol, 96% ethanol, 70% ethanol, and distilled water. The following staining procedure uses a slightly modified version of the Gallyas myelin stain (Gallyas 1971, 1979), and myelin fibers are stained black to contrast with the completely unstained background.

1. Immersion for 30 min in a solution of 80 ml pyridine and 40 ml acetic acid
2. Three rinses in distilled water (5 min each)
3. Immersion for 30 min in a solution of 1 g ammonium nitrate, 1000 ml distilled water, 1 g silver nitrate, and 3 ml 4% $NaOH$ [added in the given order]
4. Three rinses in 0.5% acetic acid (3 min each)
5. Immersion for 20–30 min in the physical developer. Four stock solutions are used to make this developer: Solution A: 50 g sodium bicarbonate in 1000 ml distilled water. Solution B: 2 g ammonium nitrate, 2 g silver nitrate, 10 g tungstosilicic acid dissolved in 1000 ml distilled water. Solution C: 2 g ammonium nitrate, 2 g silver nitrate, 10 g tungstosilicic acid, 7.3 ml 37% formaldehyde, 1000 ml distilled water. Solution D: Equal parts of distilled water and Agfa bleach fixer. The physical developer is prepared by combining 50 ml A, 15 ml B, 35 ml C, and 0.16 ml D
6. One 5-min rinse in distilled water
7. Immersion for 10 min in a photographic fixer, e.g., 20 ml Kodak Ektaflo in 70 ml distilled water [69]
8. One 5-min rinse in distilled water
9. Graded ethanol series (70%, 96% 100%)
10. Xylene bath
11. Coverslipping with Depex

Following step 8 a differentiation procedure (increase in contrast) can be carried out. This consists in a 5-min rinse in 0.5% acetic acid, a 15- to 30-s differentiating bath (bleach fixer as in solution D), and two rinses in 0.5% acetic acid.

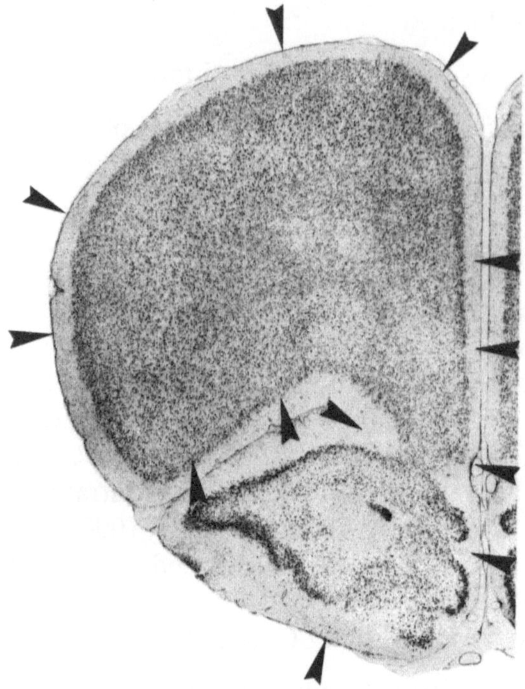

a

[70]

Fig. 55 a–d

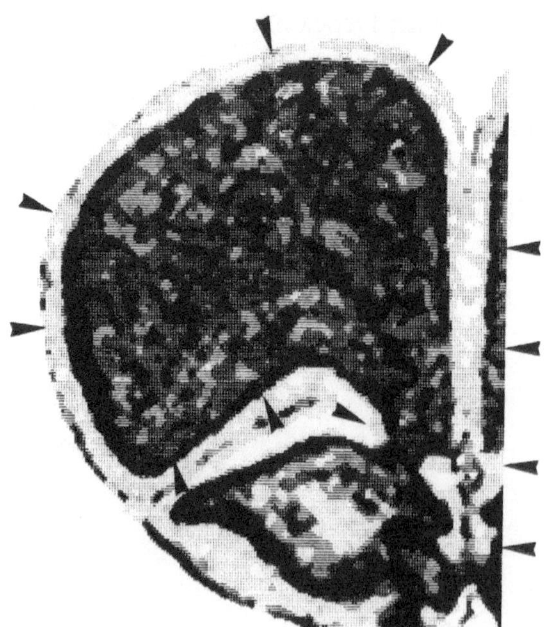

b

c

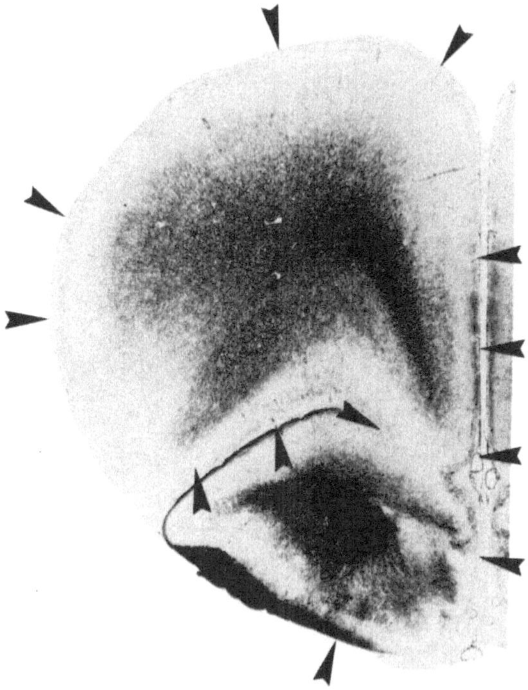

d

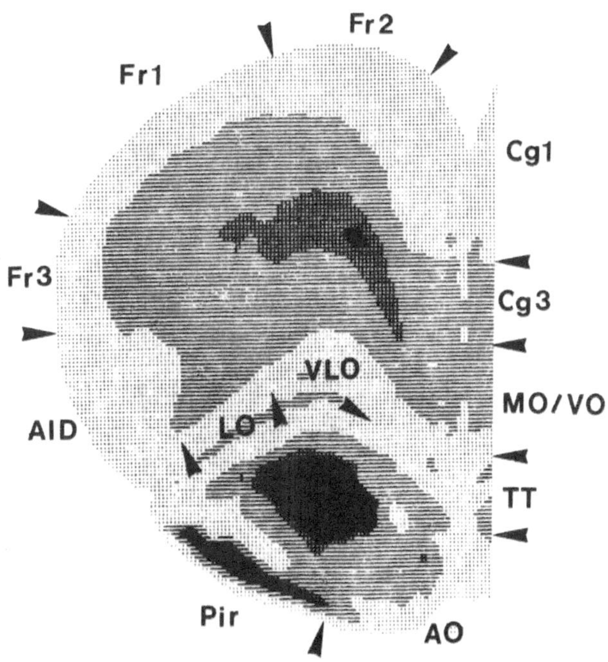

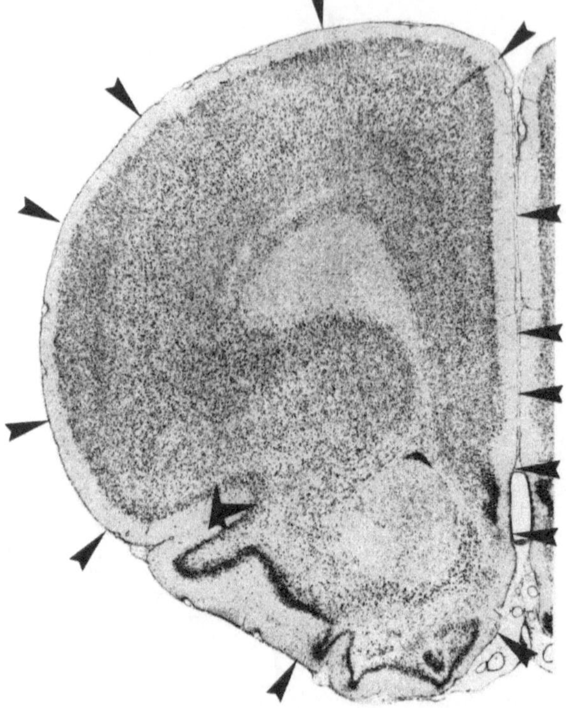

a

[72]

Fig. 56 a–d

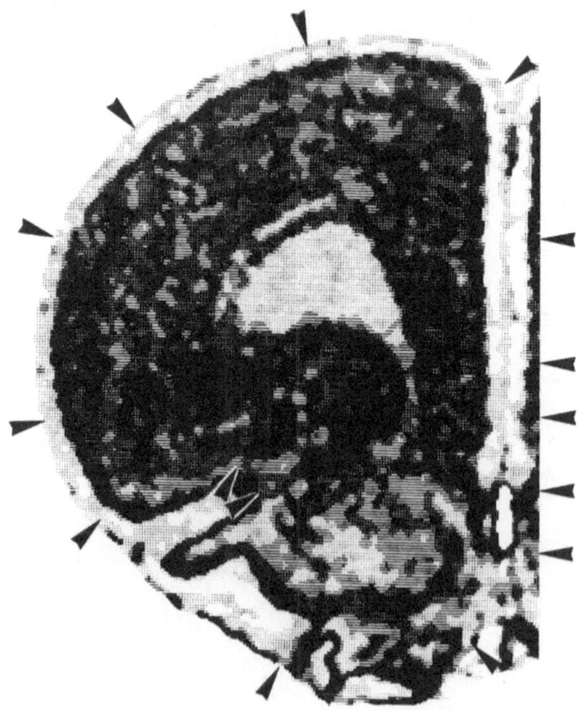

b

c

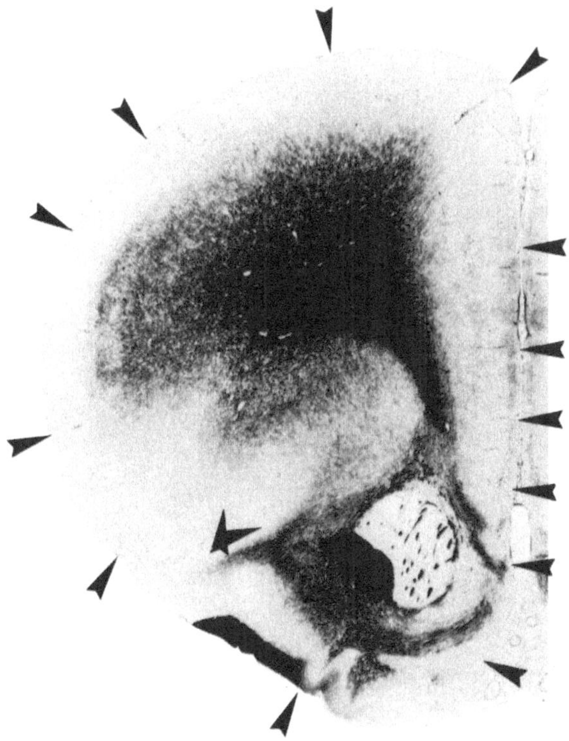

d

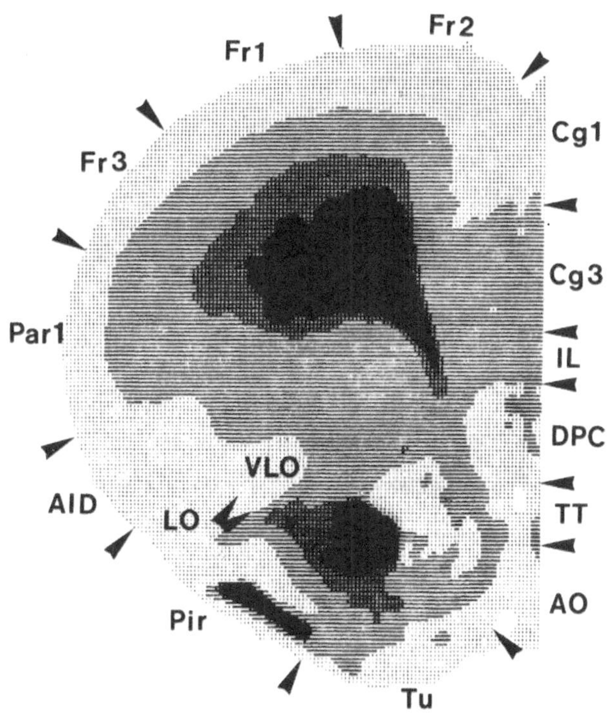

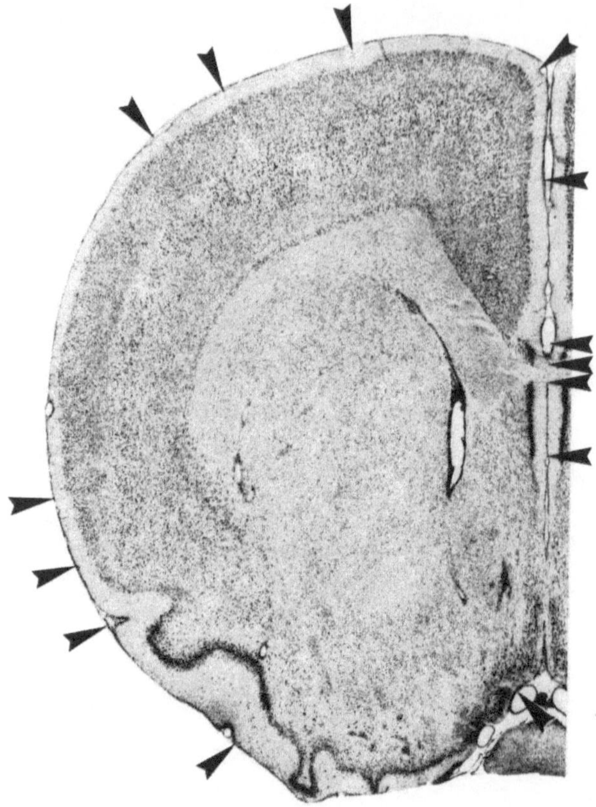

a

[74]

Fig. 57 a–d

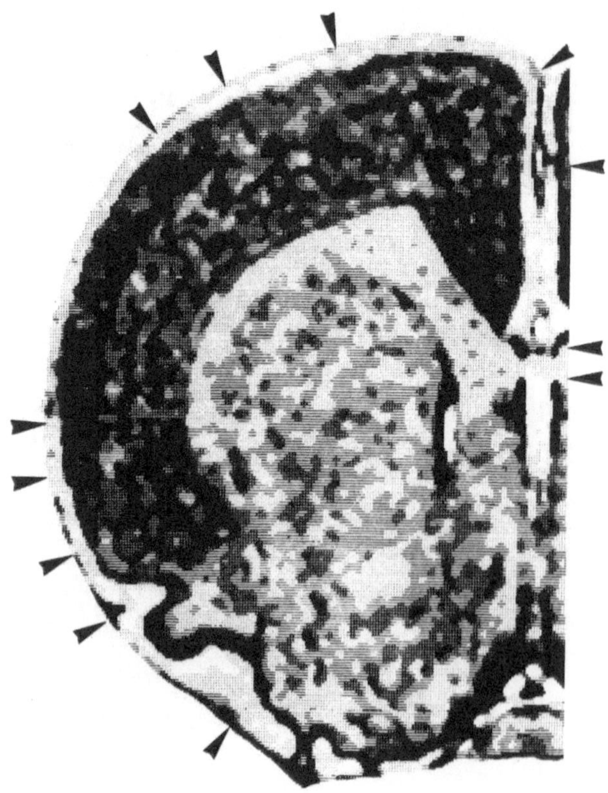

b

c

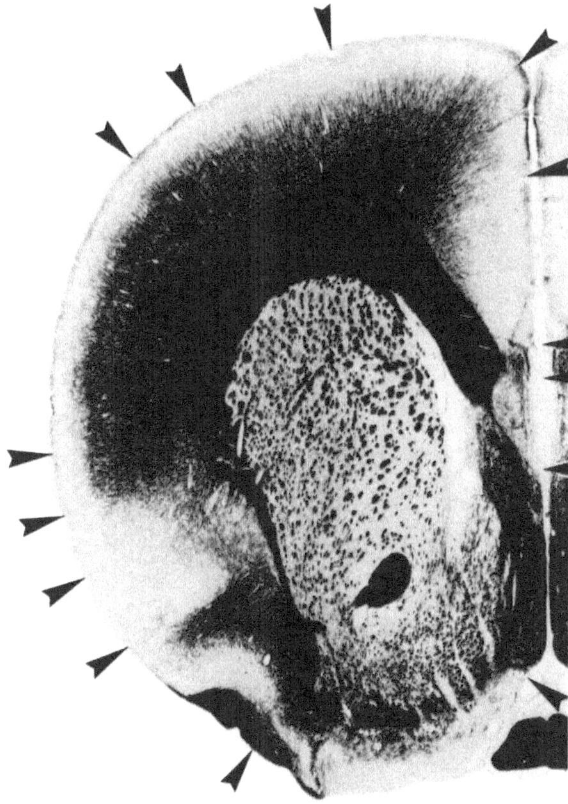

d

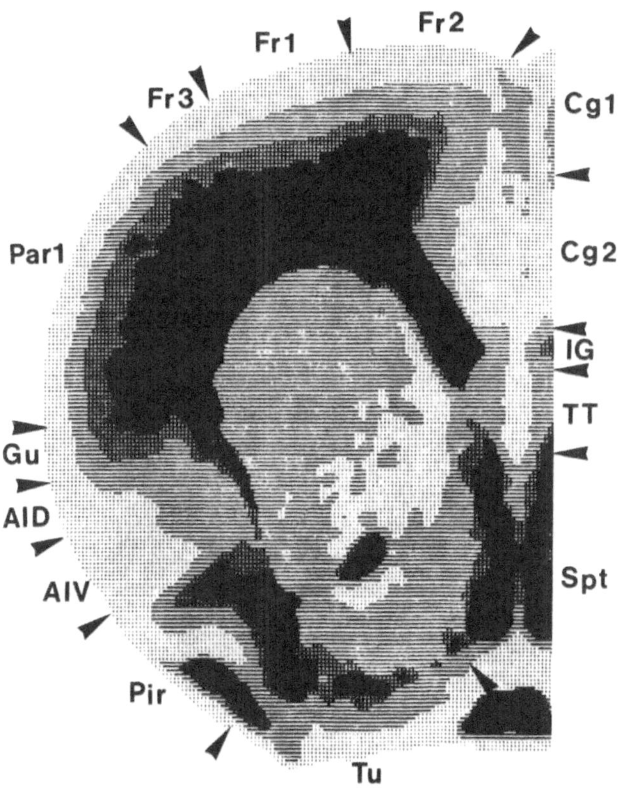

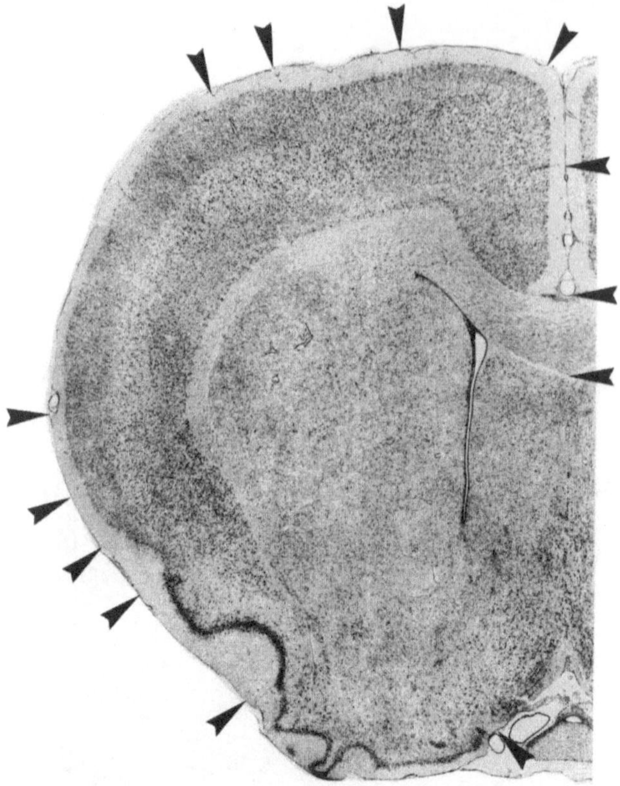

a

Fig. 58a–d

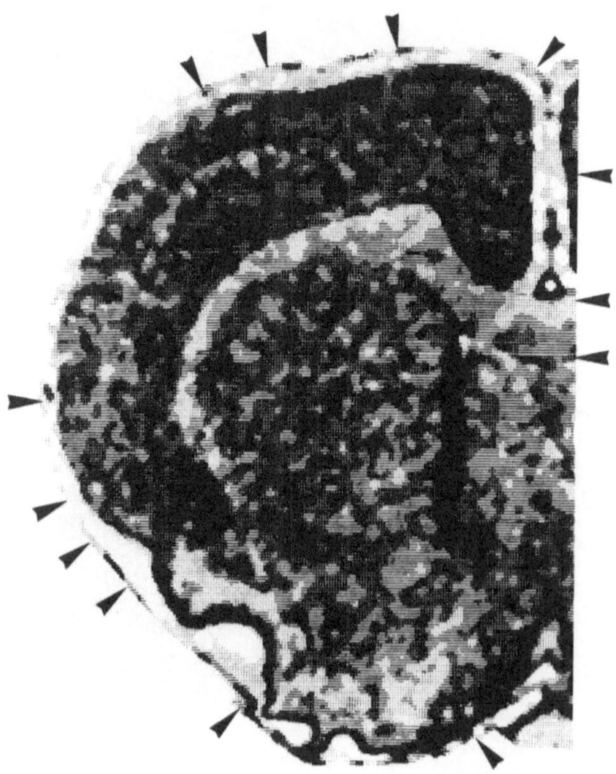

b

c

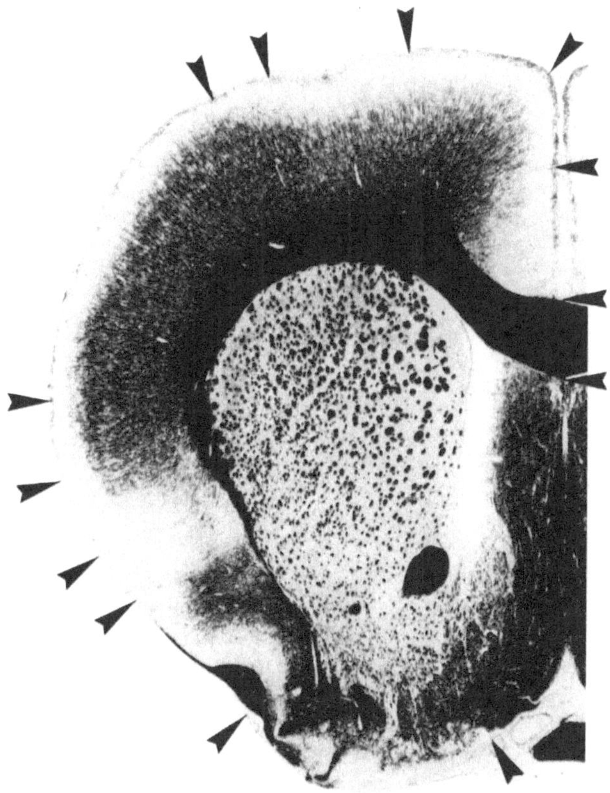

d

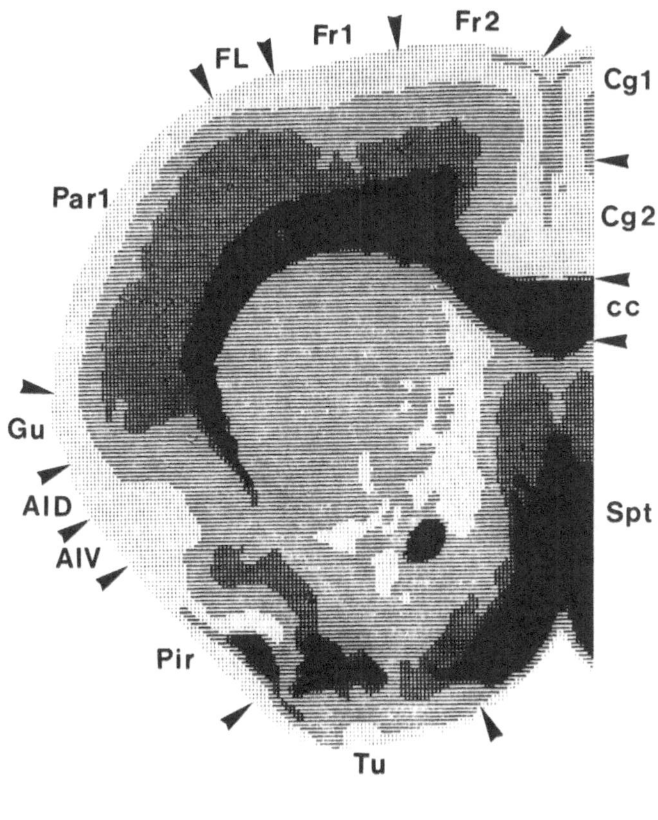

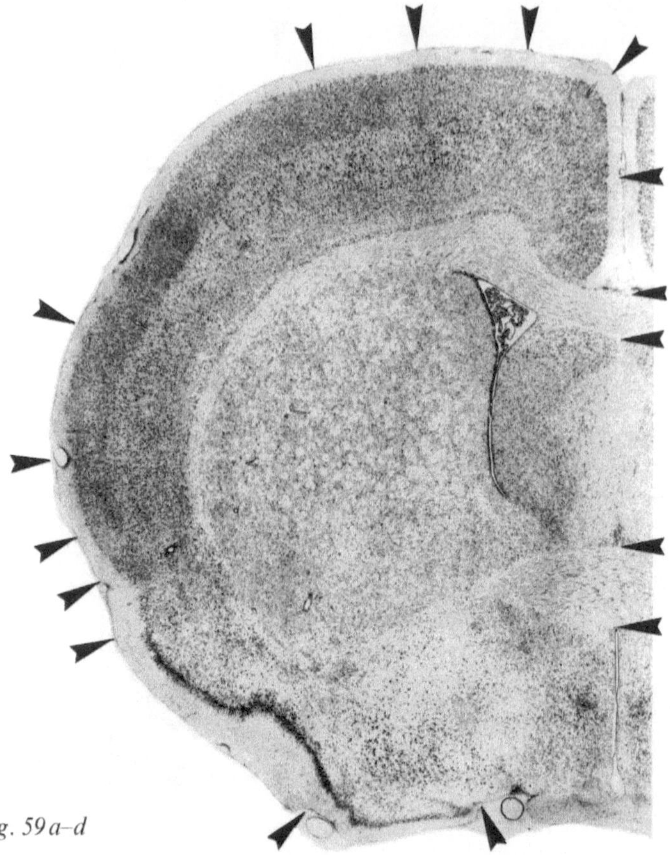

a

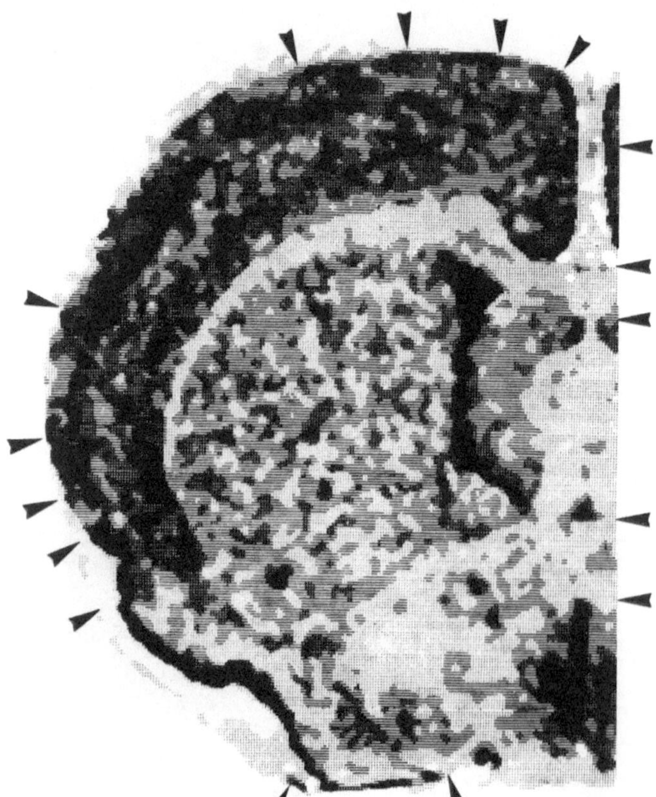

b

Fig. 59a–d

c

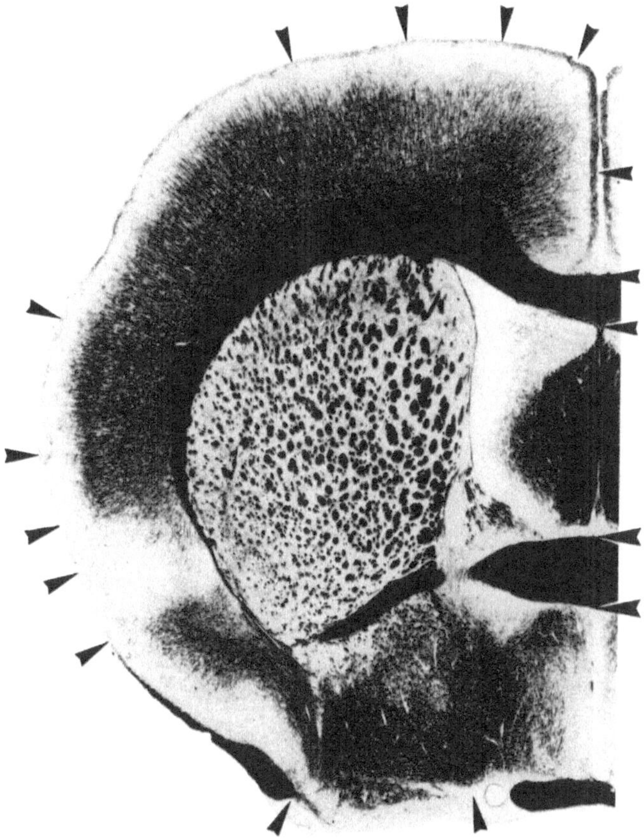

d

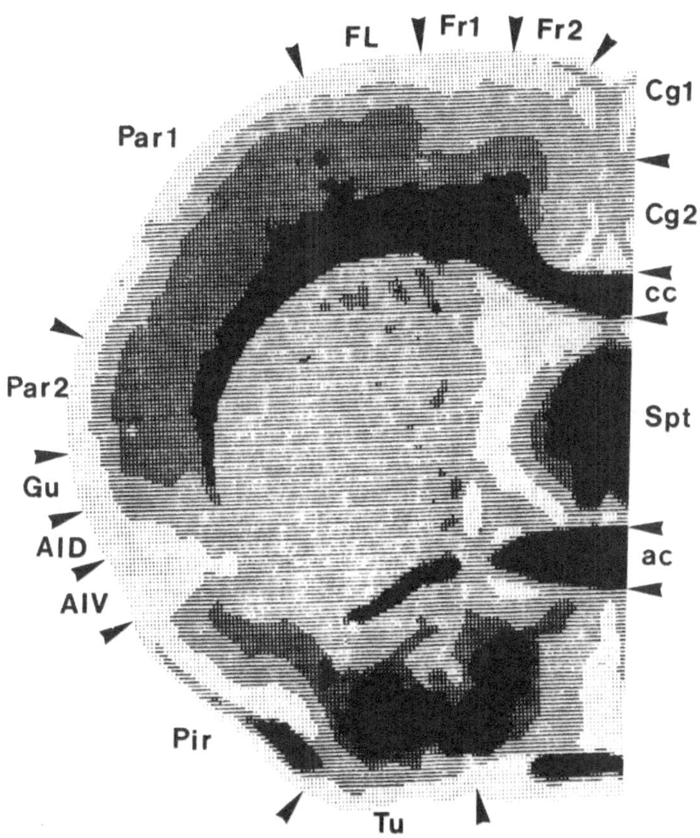

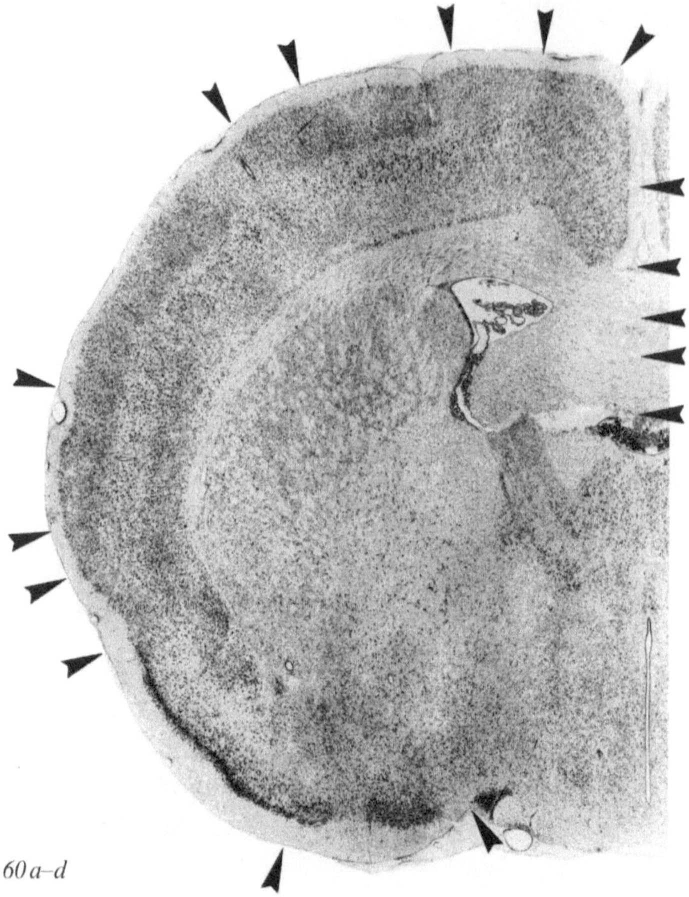

a

Fig. 60 a–d

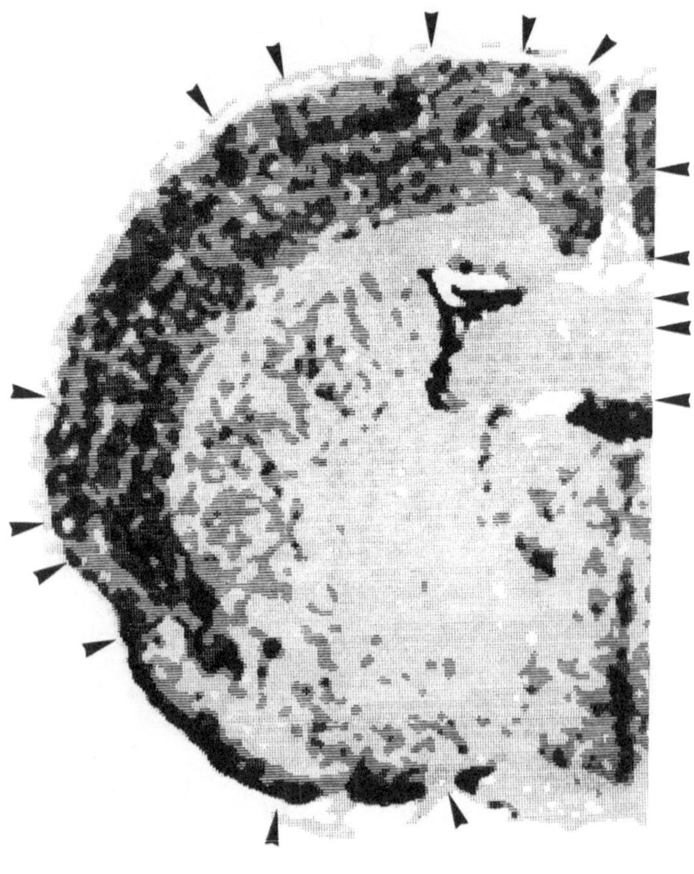

b

c

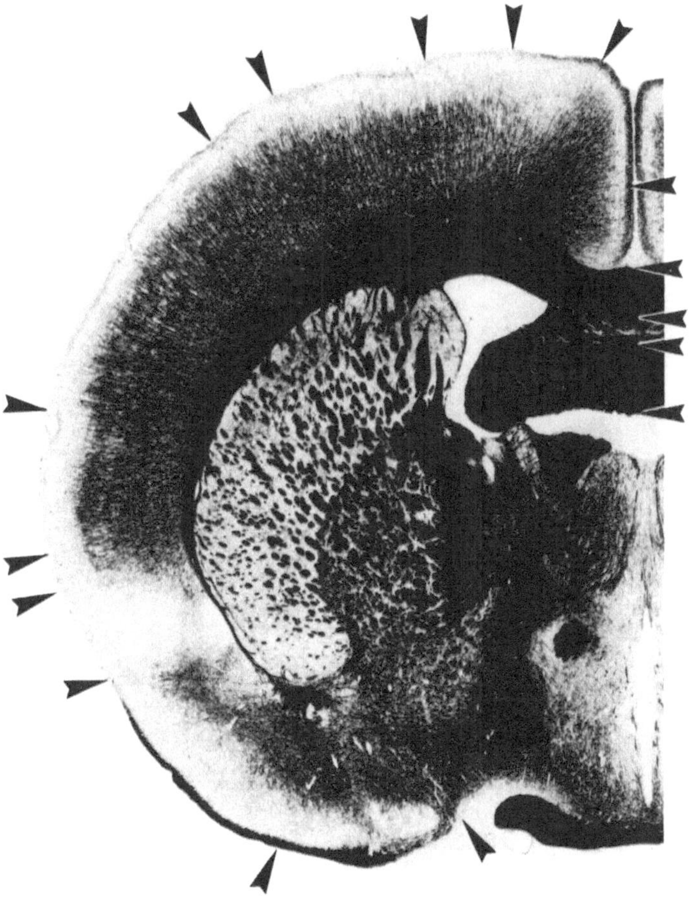

[81]

d

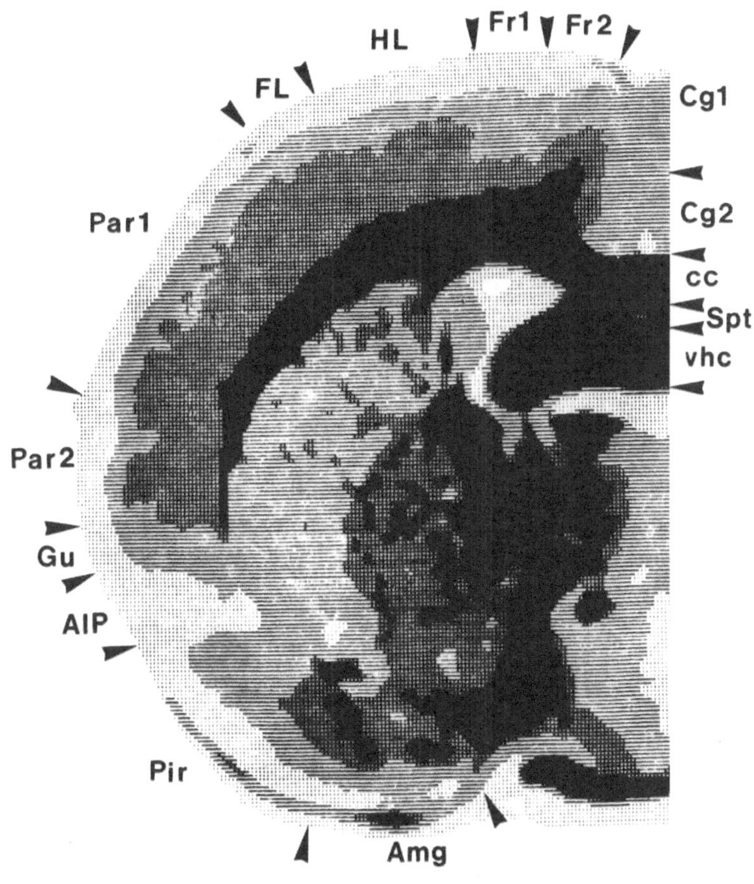

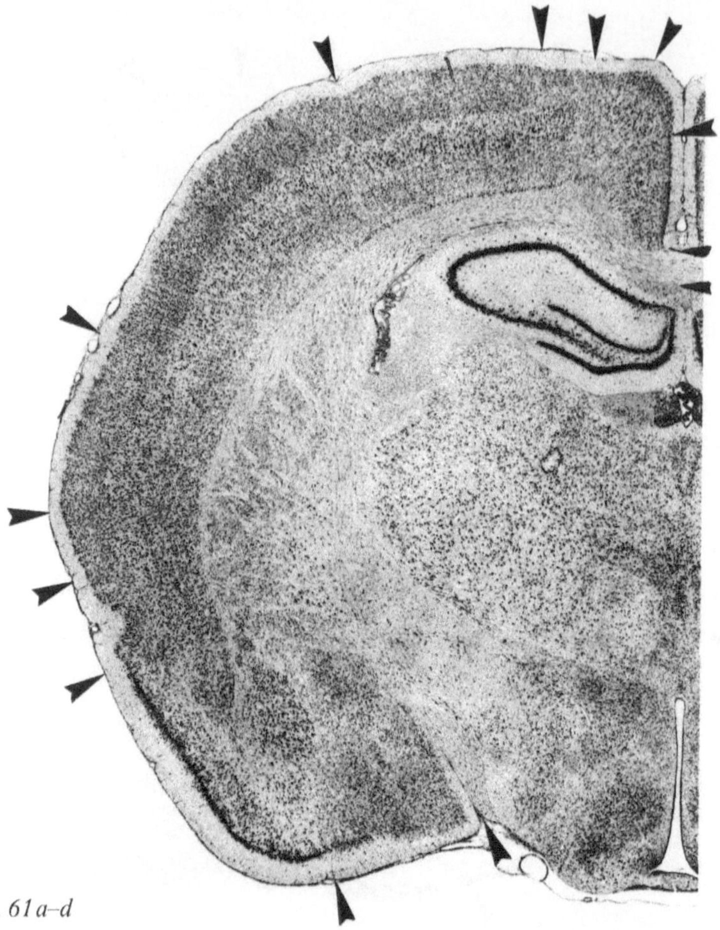

a

Fig. 61 a–d

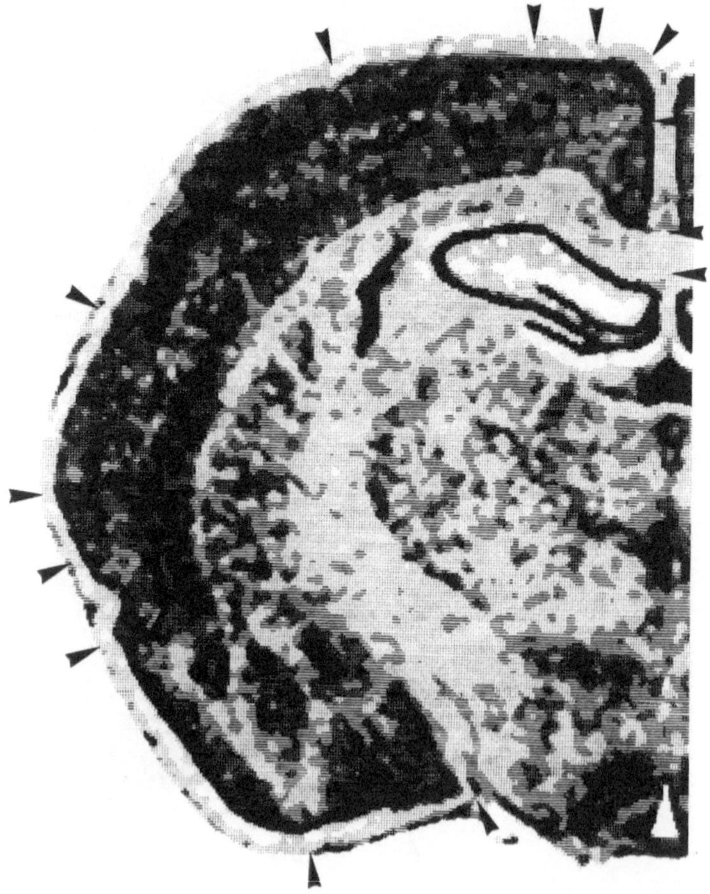

b

c

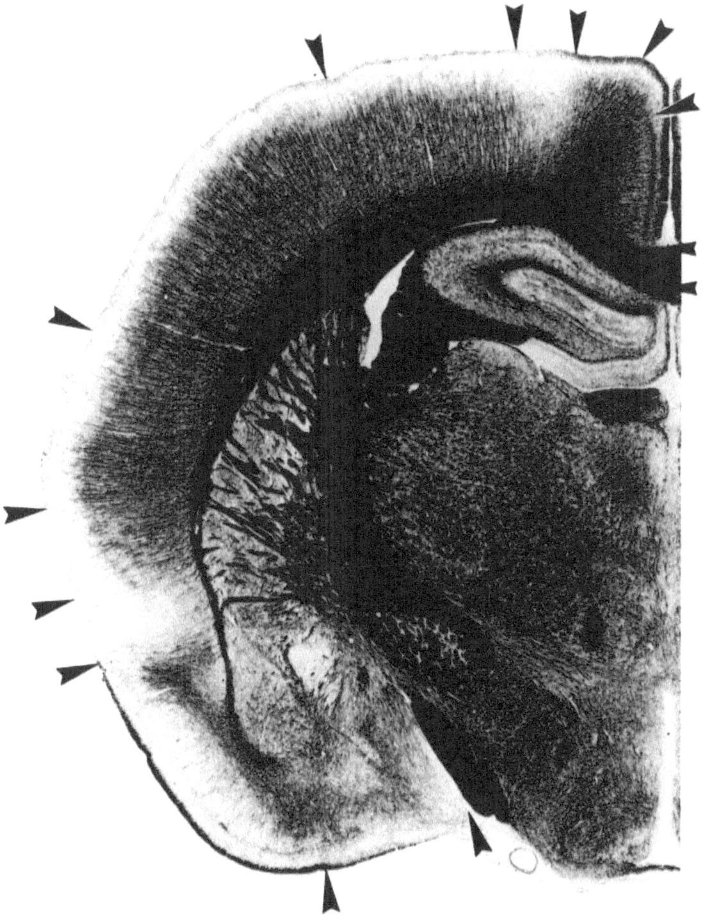

d

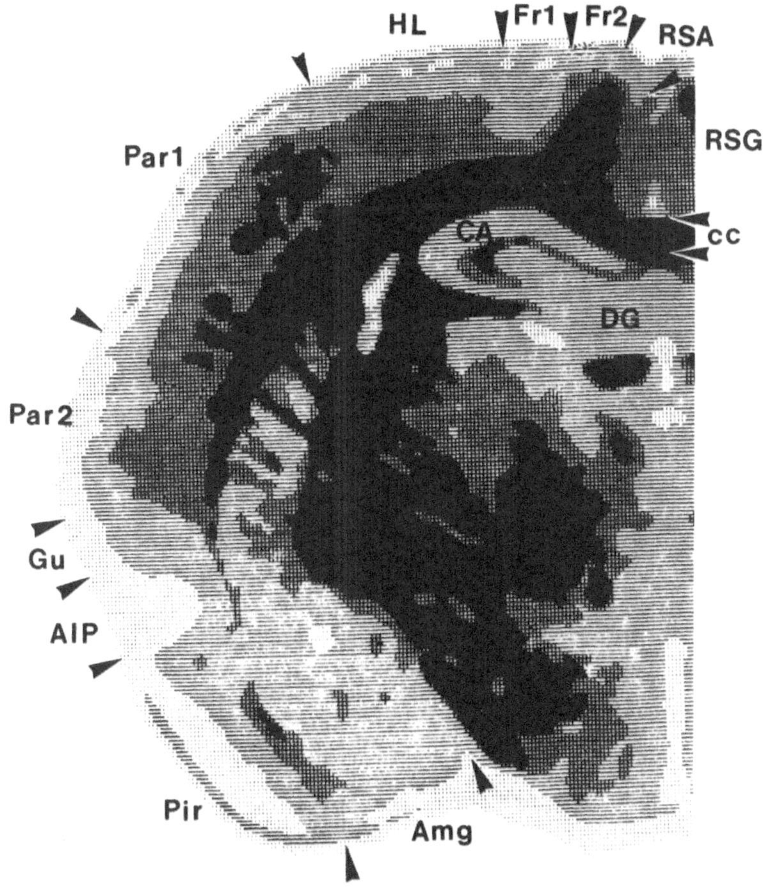

[84]

Fig. 62a–d

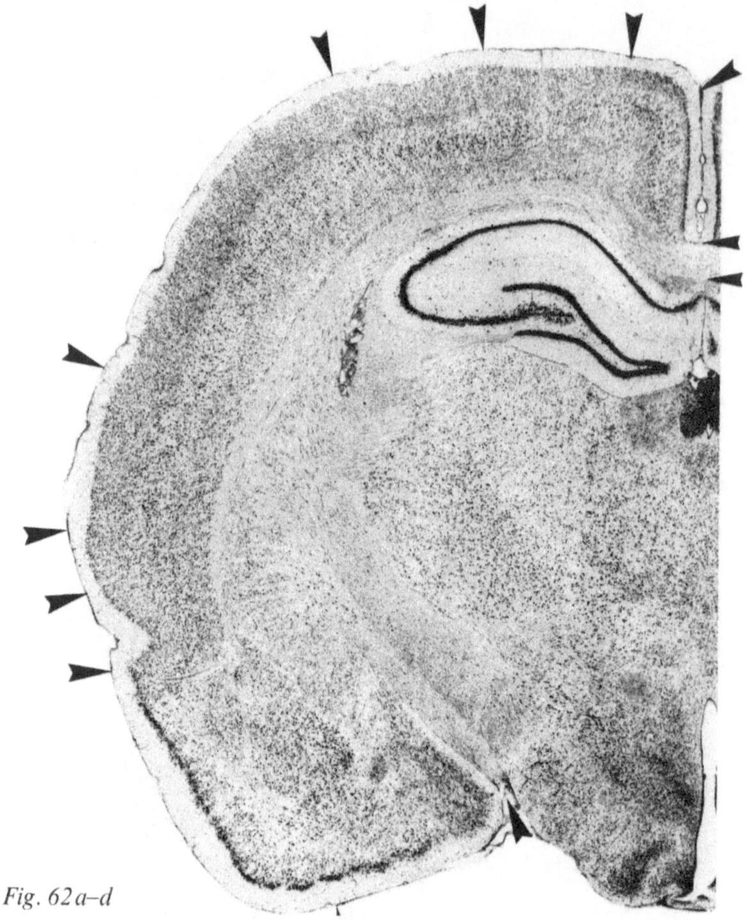

a

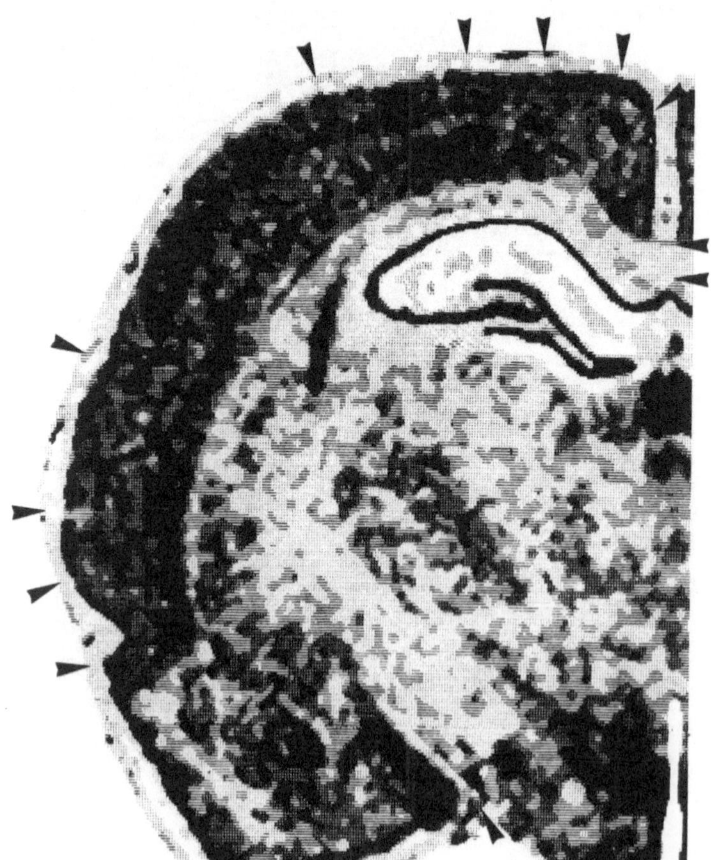

b

c

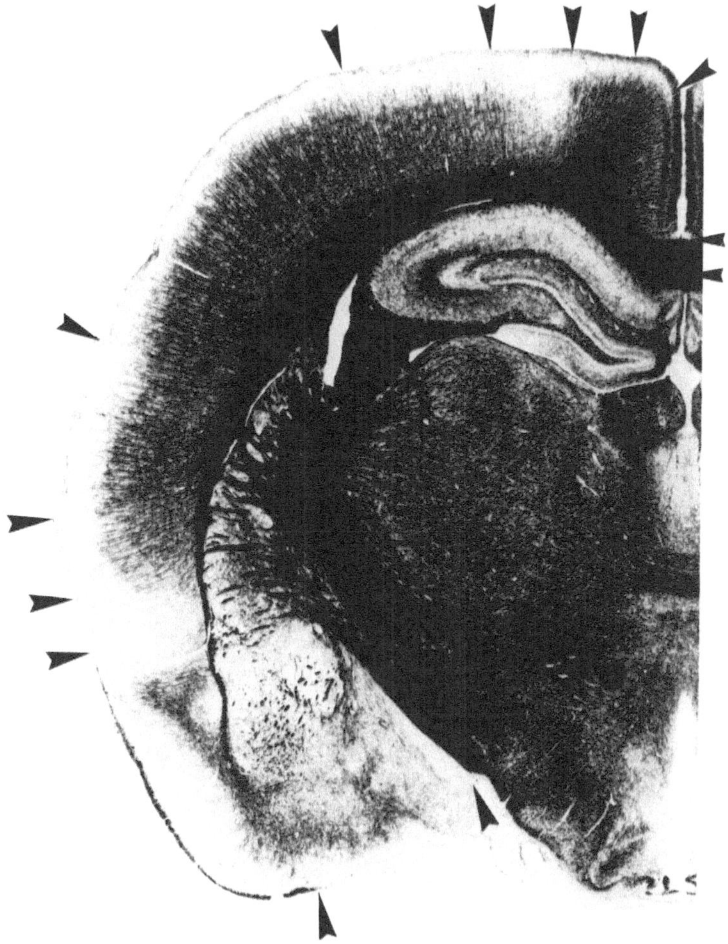

d

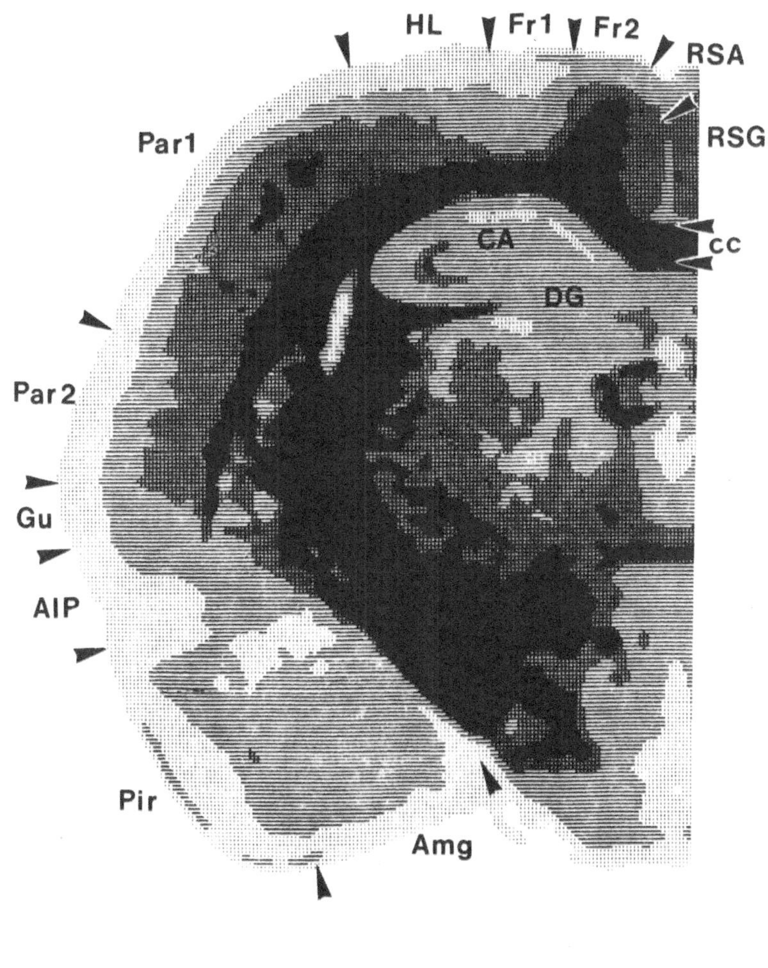

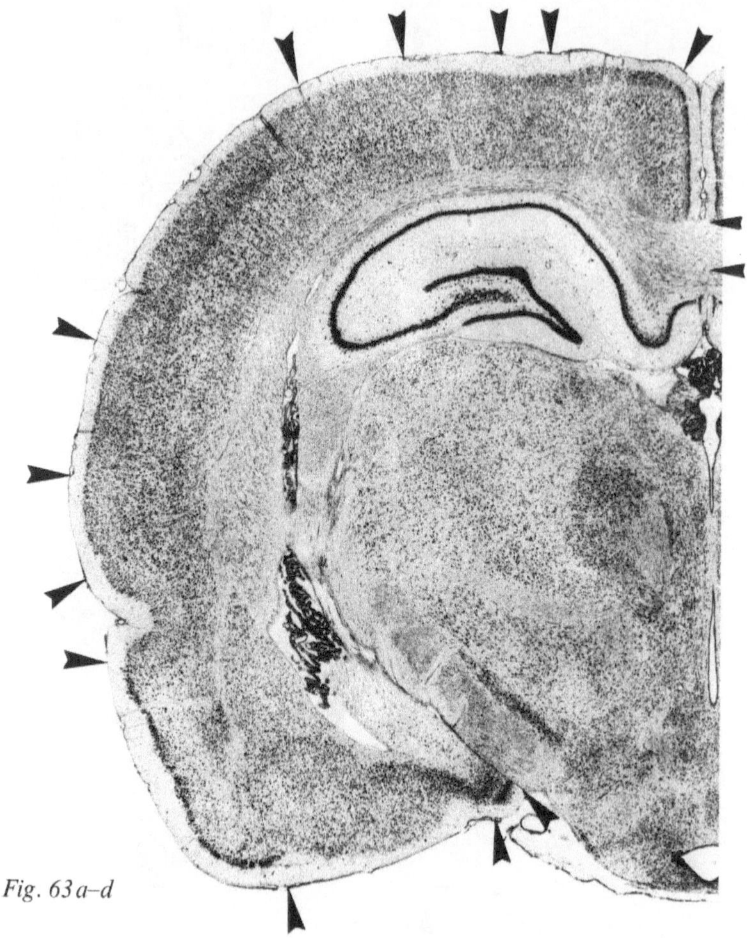

a

Fig. 63 a–d

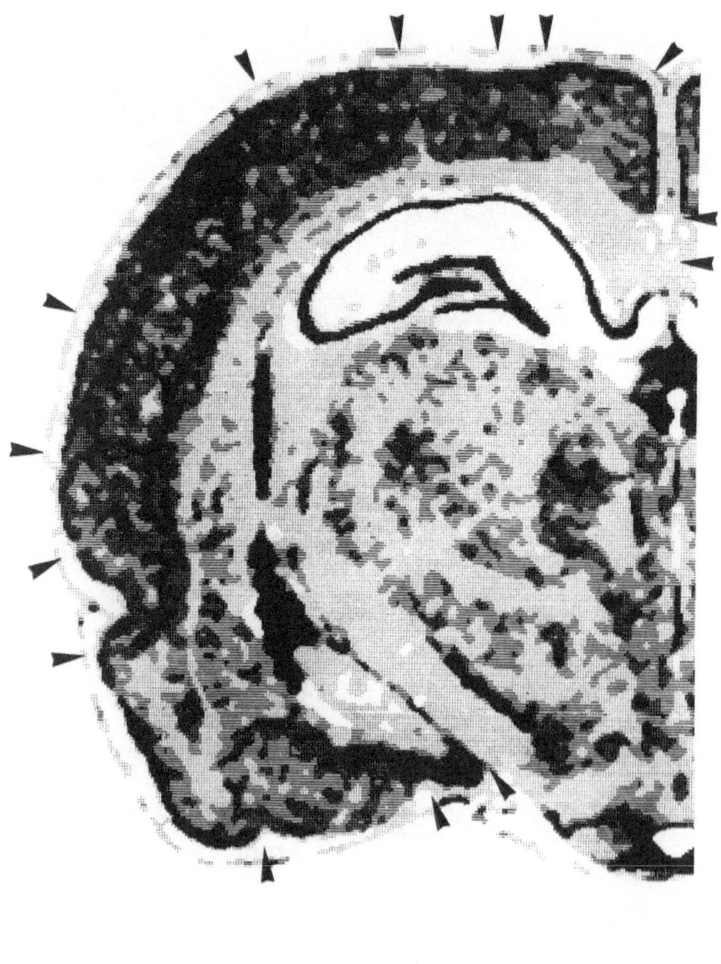

b

c

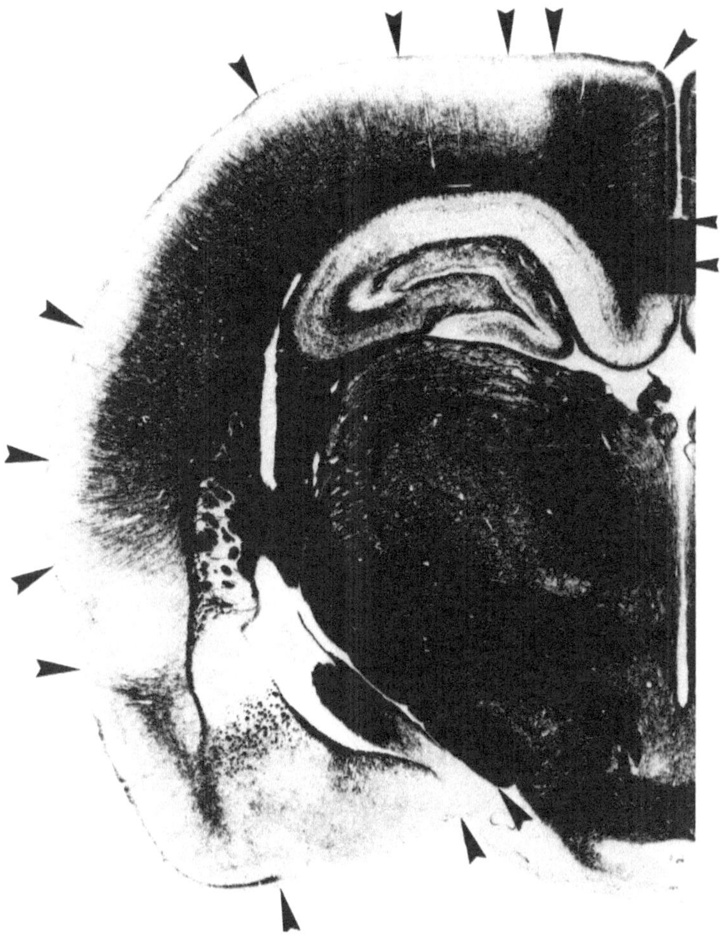

d

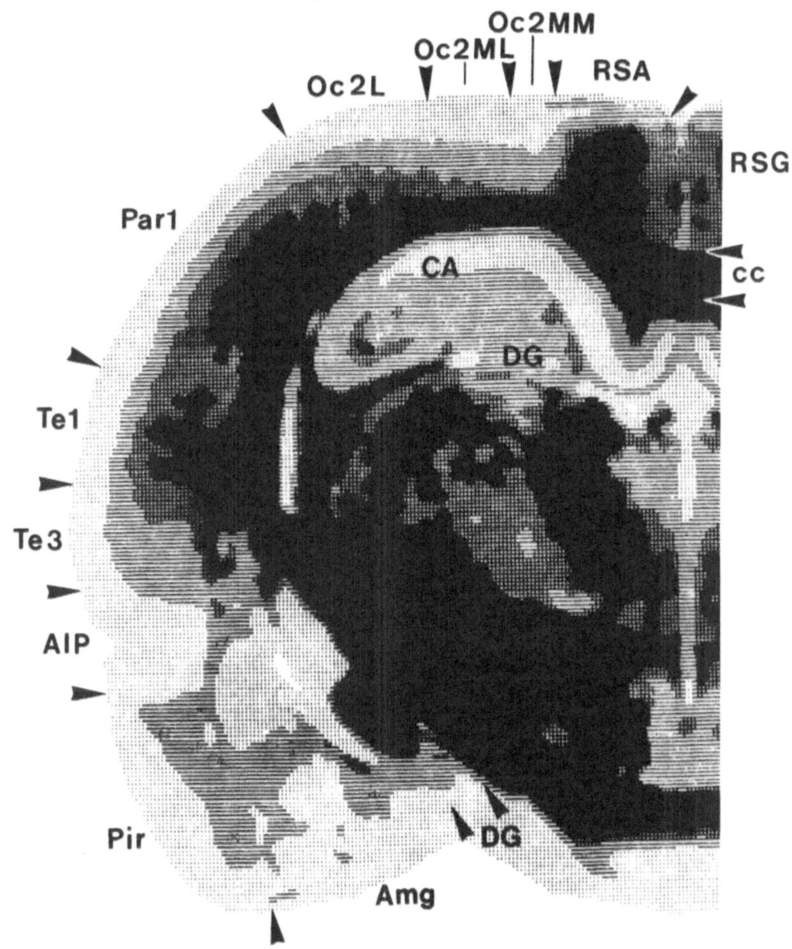

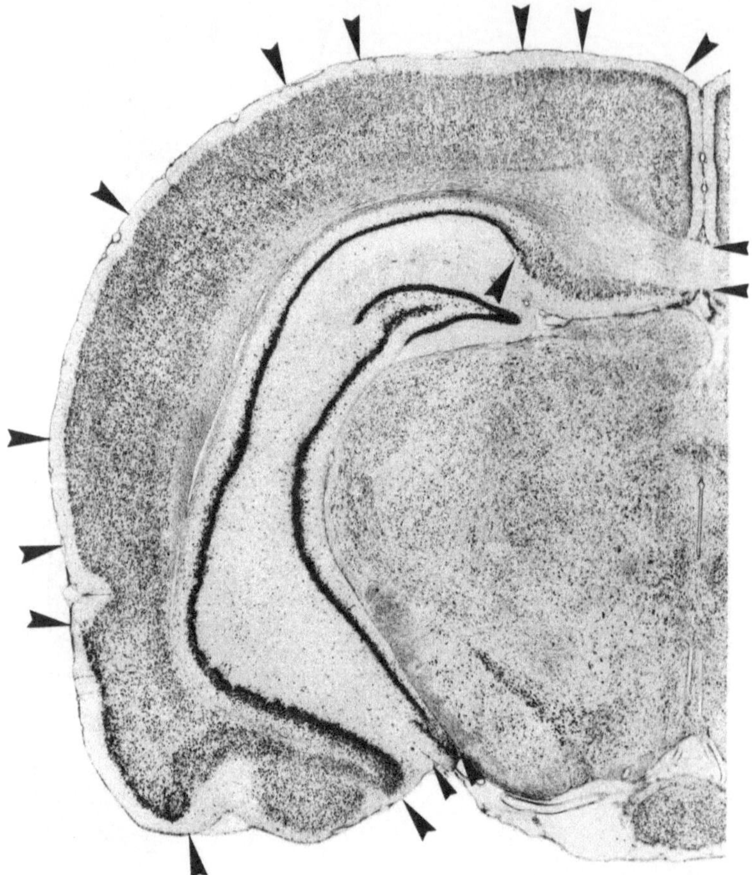

a

Fig. 64 a–d

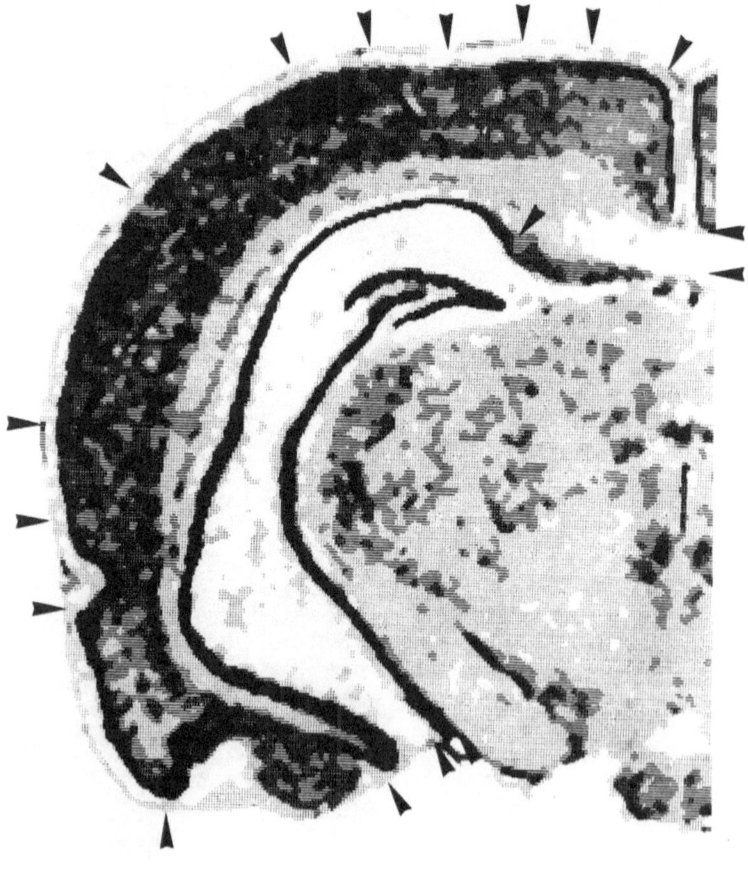

b

c

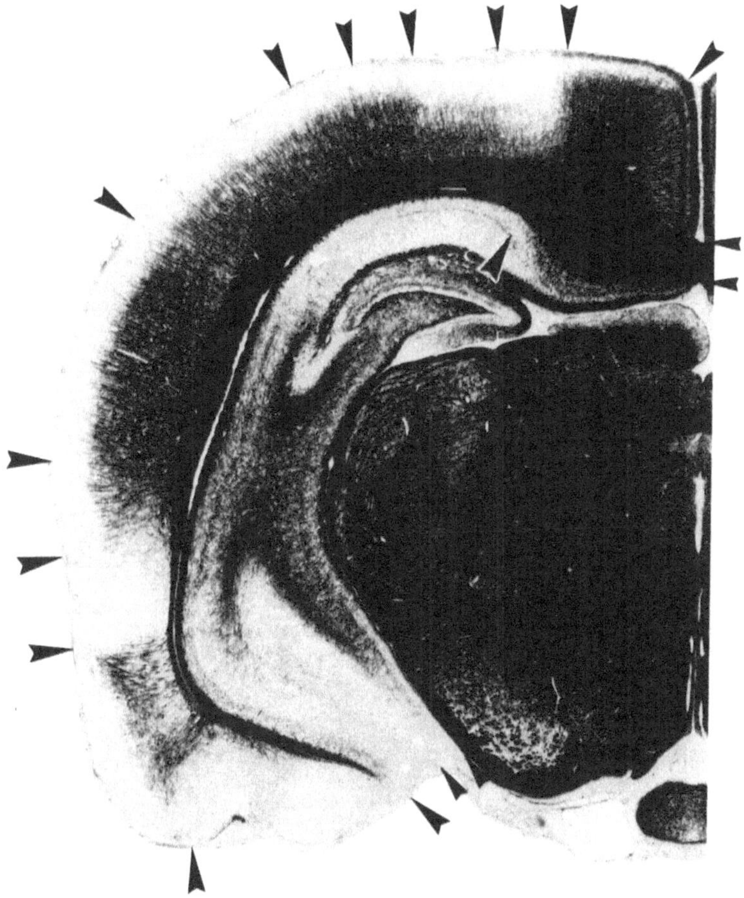

d

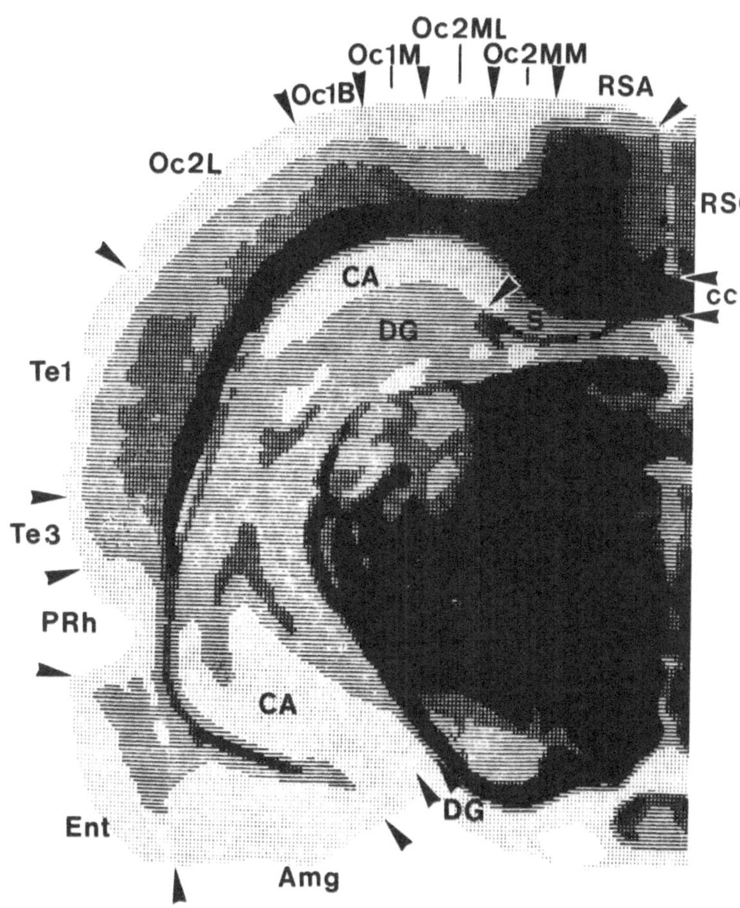
Oc2ML
Oc1M
Oc2MM
Oc1B
RSA
Oc2L
RSG
CA
cc
DG
S
Te1
Te3
PRh
CA
DG
Ent
Amg

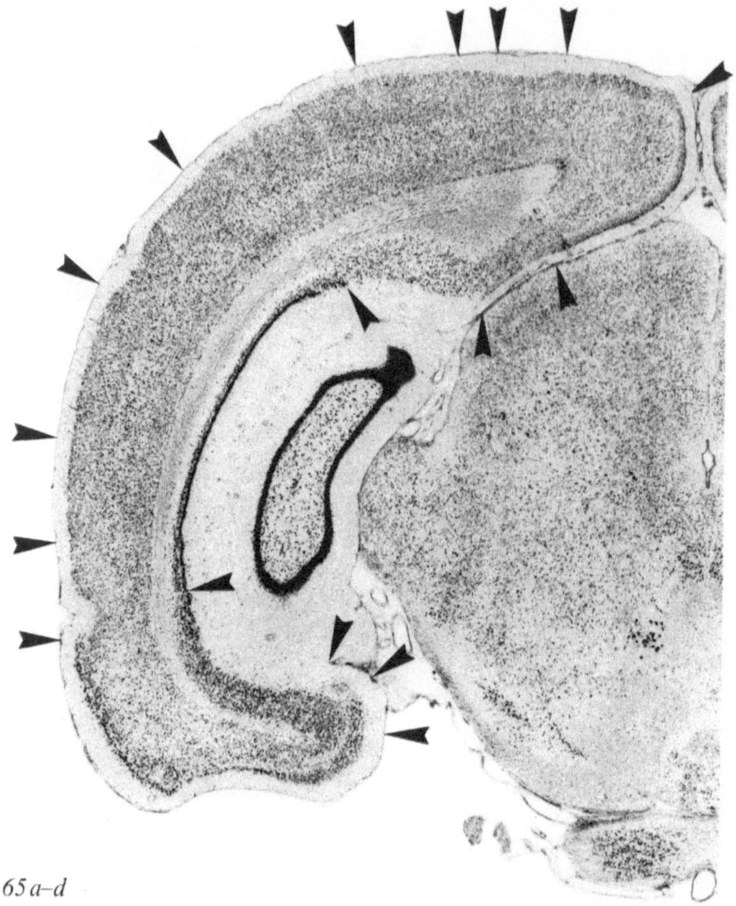

a

Fig. 65 a–d

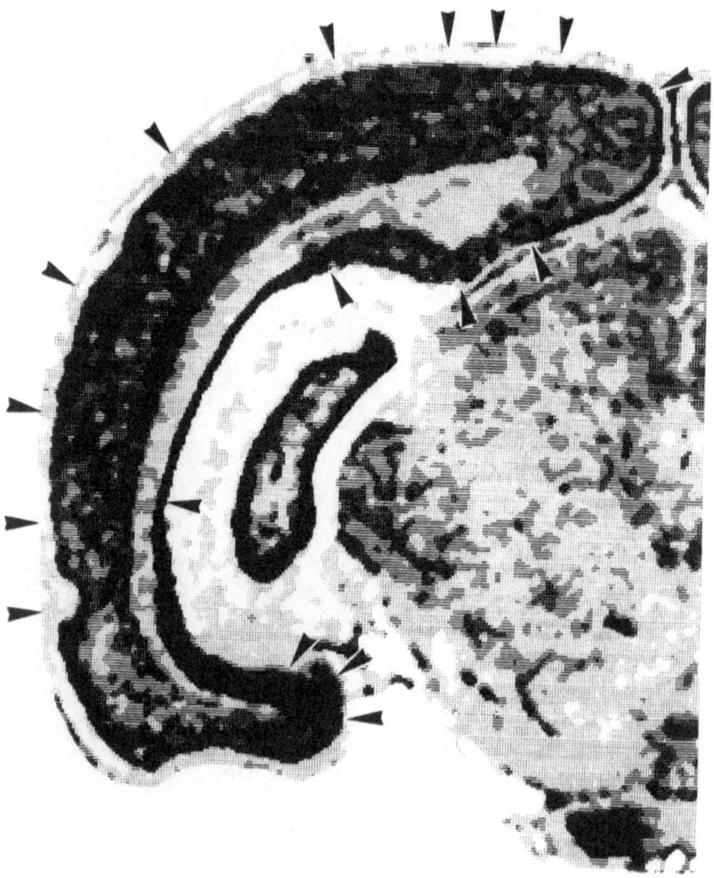

b

c

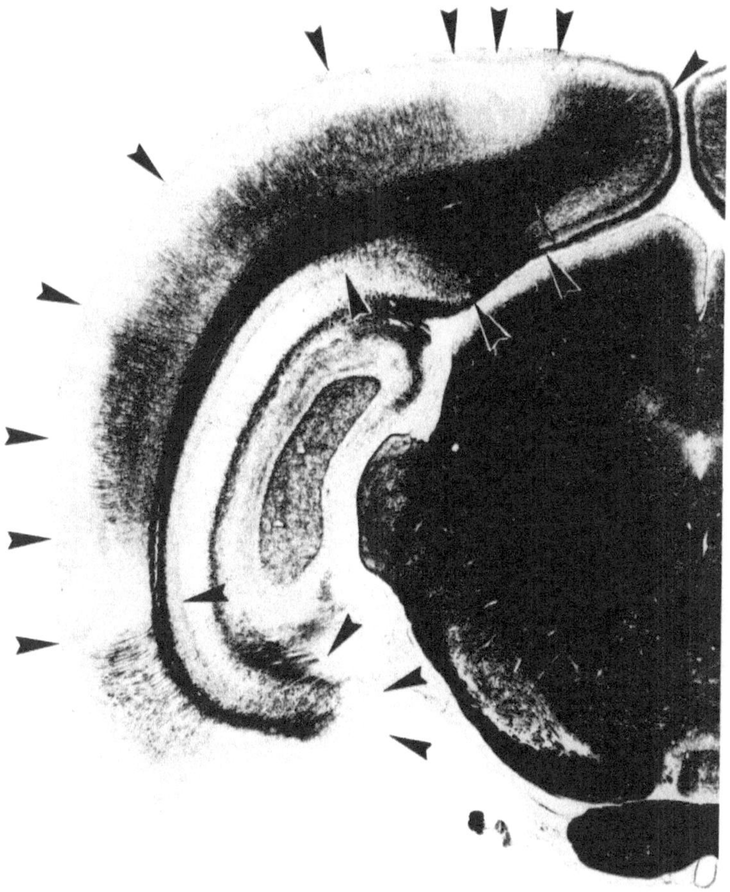

d

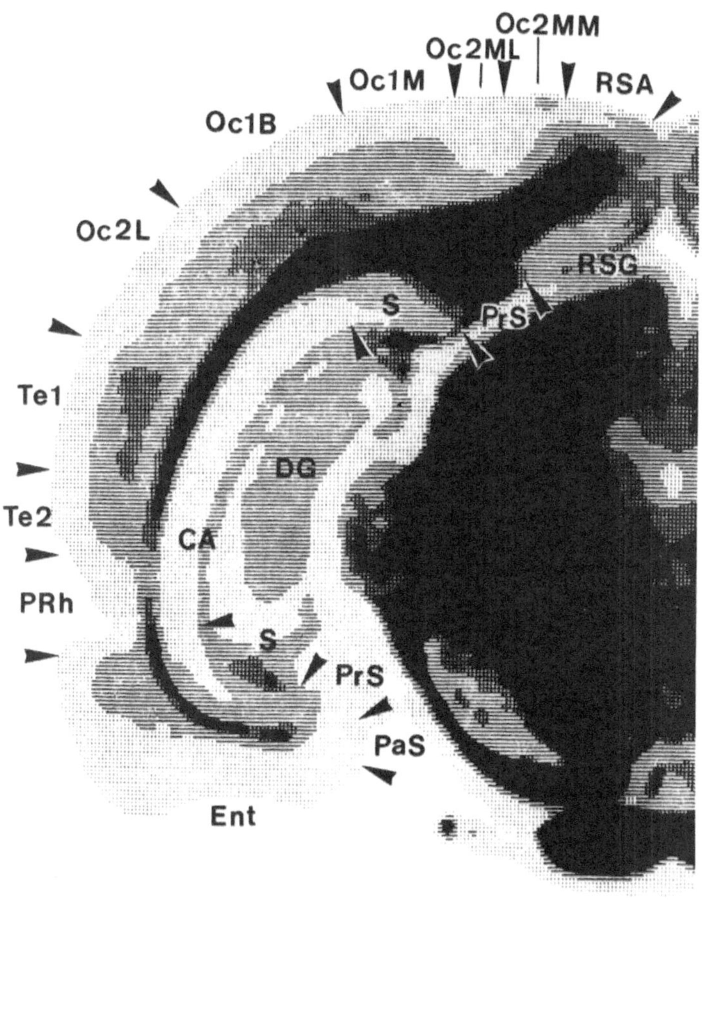
Oc2MM
Oc2ML
Oc1M
RSA
Oc1B
Oc2L
RSG
S
PrS
Te1
DG
Te2
CA
PRh
S
PrS
PaS
Ent

Fig. 66 a–d

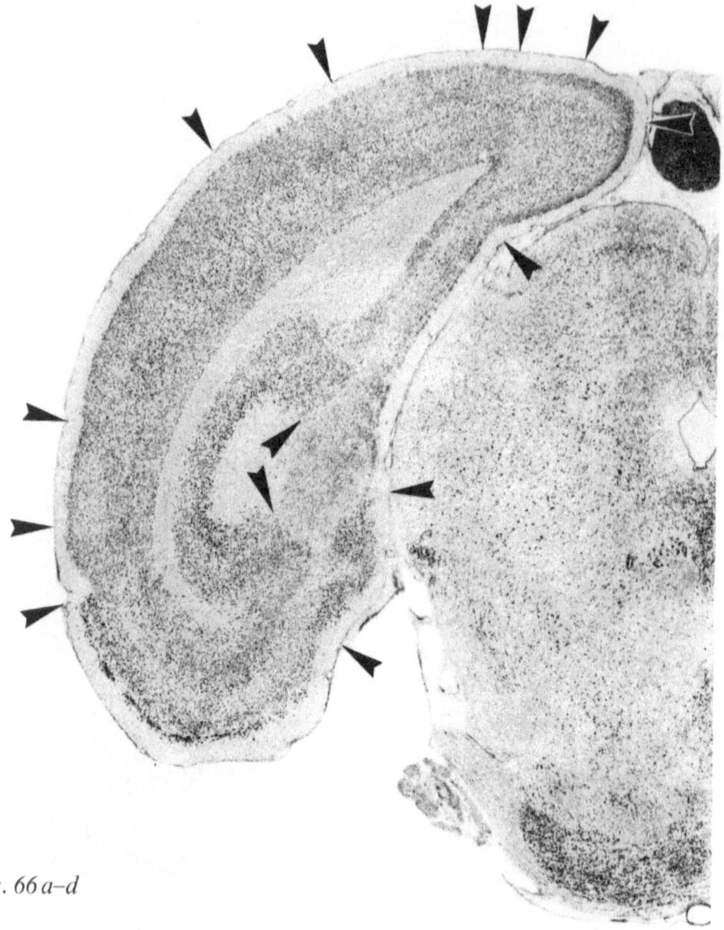

a

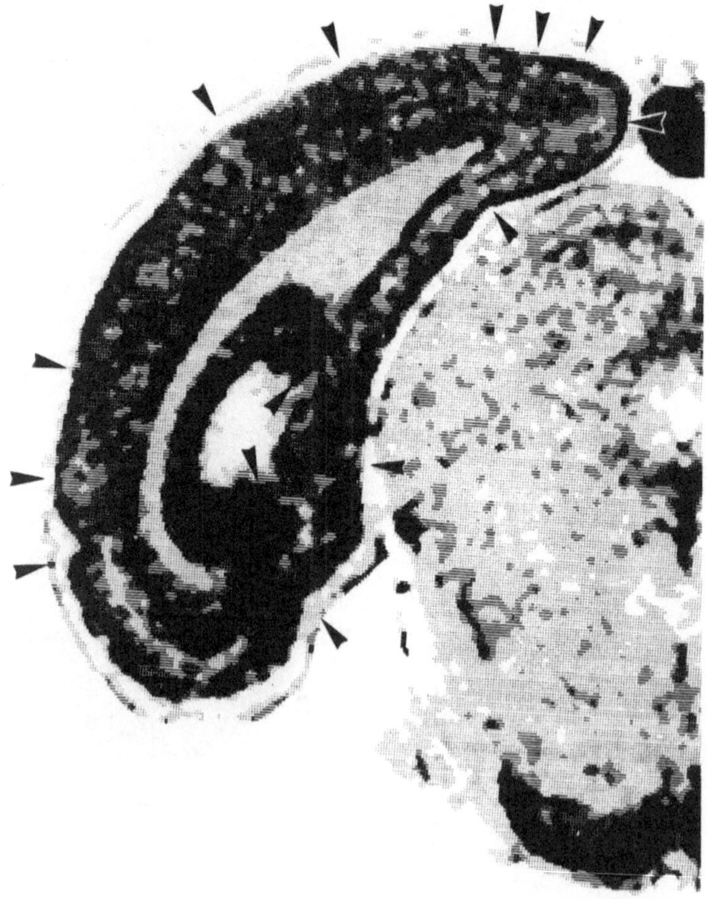

b

c

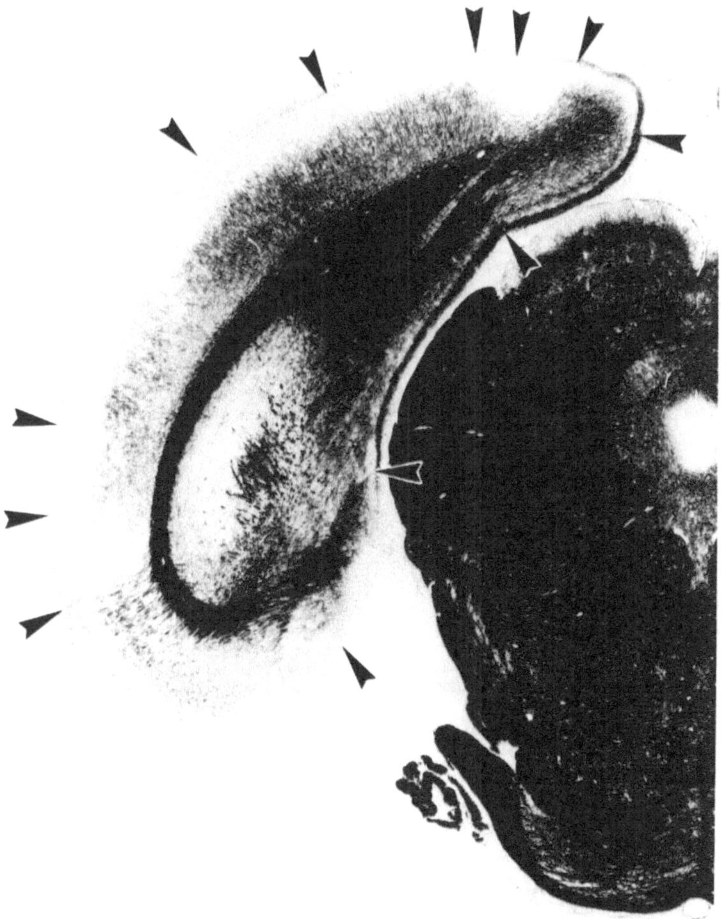

d

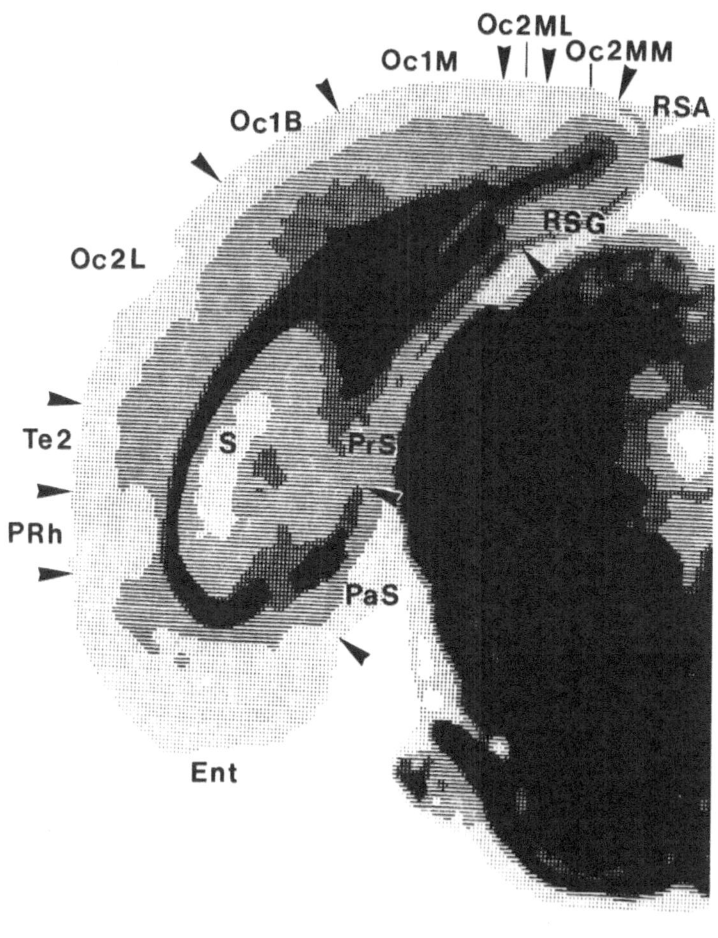

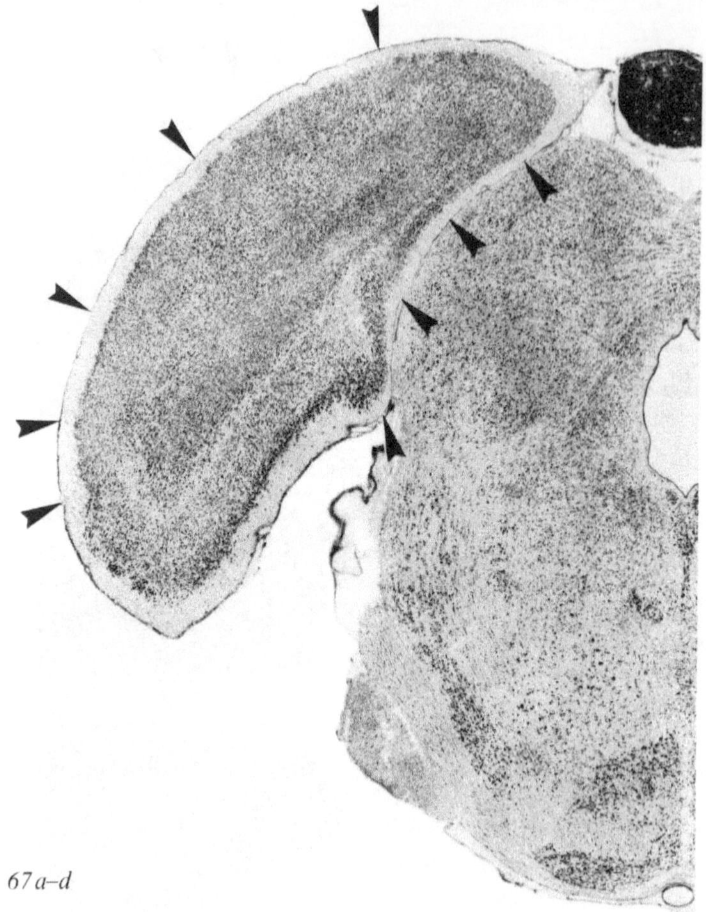

a

Fig. 67a–d

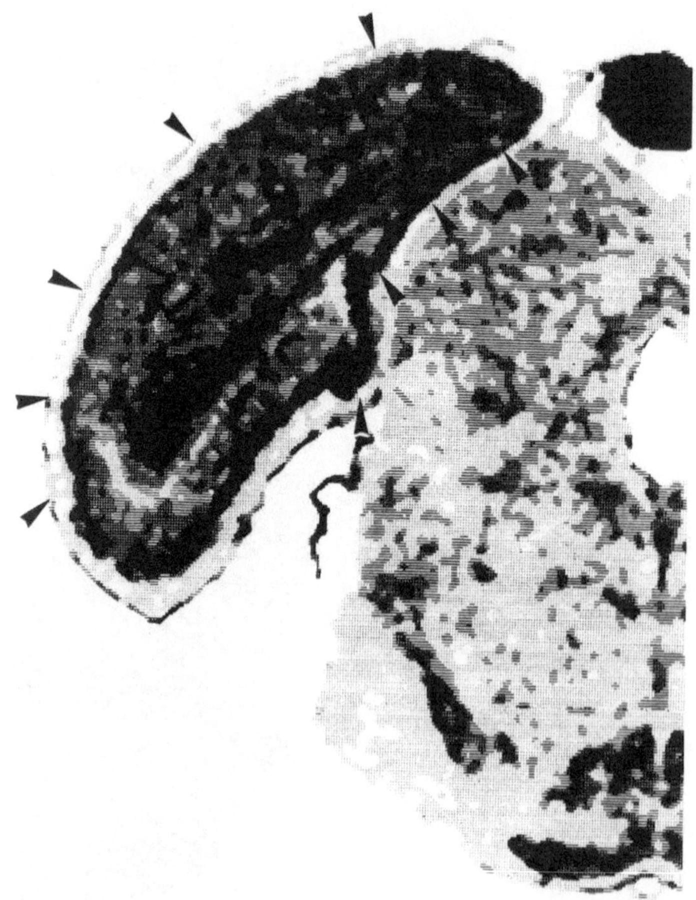

b

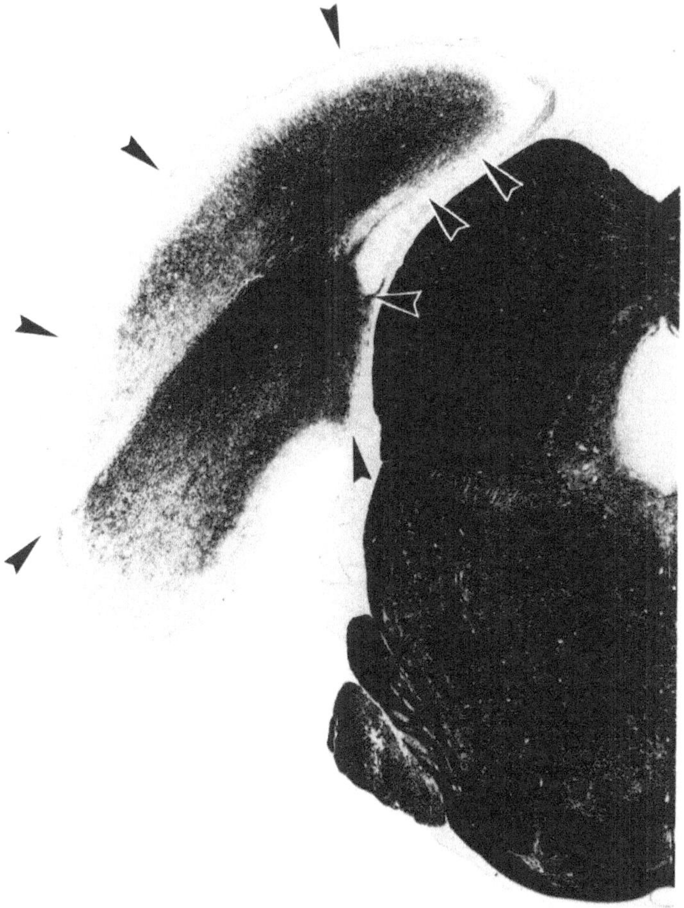

c

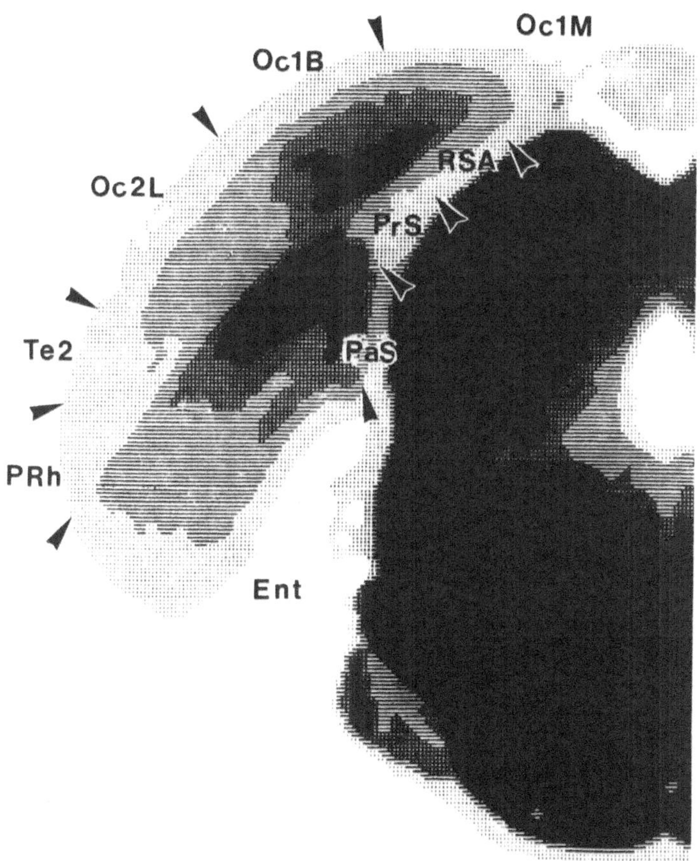

d

Laminar Structure of Cortical Areas in the Nissl- and Myelin-Stained Sections

Boundaries of cortical areas (Figs. 68–72) were determined in Nissl- and myelin-stained sections by the quantitative method described above (p. 6). In most cases the areal borders delineated in Nissl- and myelin-stained specimens from the same plane coincided precisely. In a few cases the areal boundaries shown are based on analysis by only one of the methods.

This atlas presents the stereotaxic position, the laminar pattern, and the Nissl and myelin features of those cortical areas and subfields which are often discussed in morphological, physiological and psychological studies, as revealed by Nissl and myelin staining.

AChE-stained sections of the primary visual cortex have been included because this histochemical method is a valuable tool for definition of the extension of lamina IV of the primary visual cortex (Zilles et al. 1984). This easily recognizable feature agrees with experiments involving transneuronal transport of intraocularly injected ^{3}H-proline (Zilles et al. 1984).

Fig. 68a–d. Laminar pattern of cortical areas in the frontal cortex. Areas Fr1 (*a*), Fr2 (*b*) and Fr3 (*c*) are shown at bregma 2.7 (cf. Fig. 7) and areas Cg1 and Cg2 (*d*) at bregma −0.3 (cf. Fig. 13). Nissl; 64×

Fig. 69a–d. Laminar pattern of cortical areas in the parietal cortex. Area Par1 (*a*) is shown at bregma 1.7 (cf. Fig. 9), areas Par2 (*b*) and HL (*d*) at bregma −3.3 (cf. Fig. 19) and area FL (*c*) at bregma −0.3 (cf. Fig. 13). Nissl; 64×

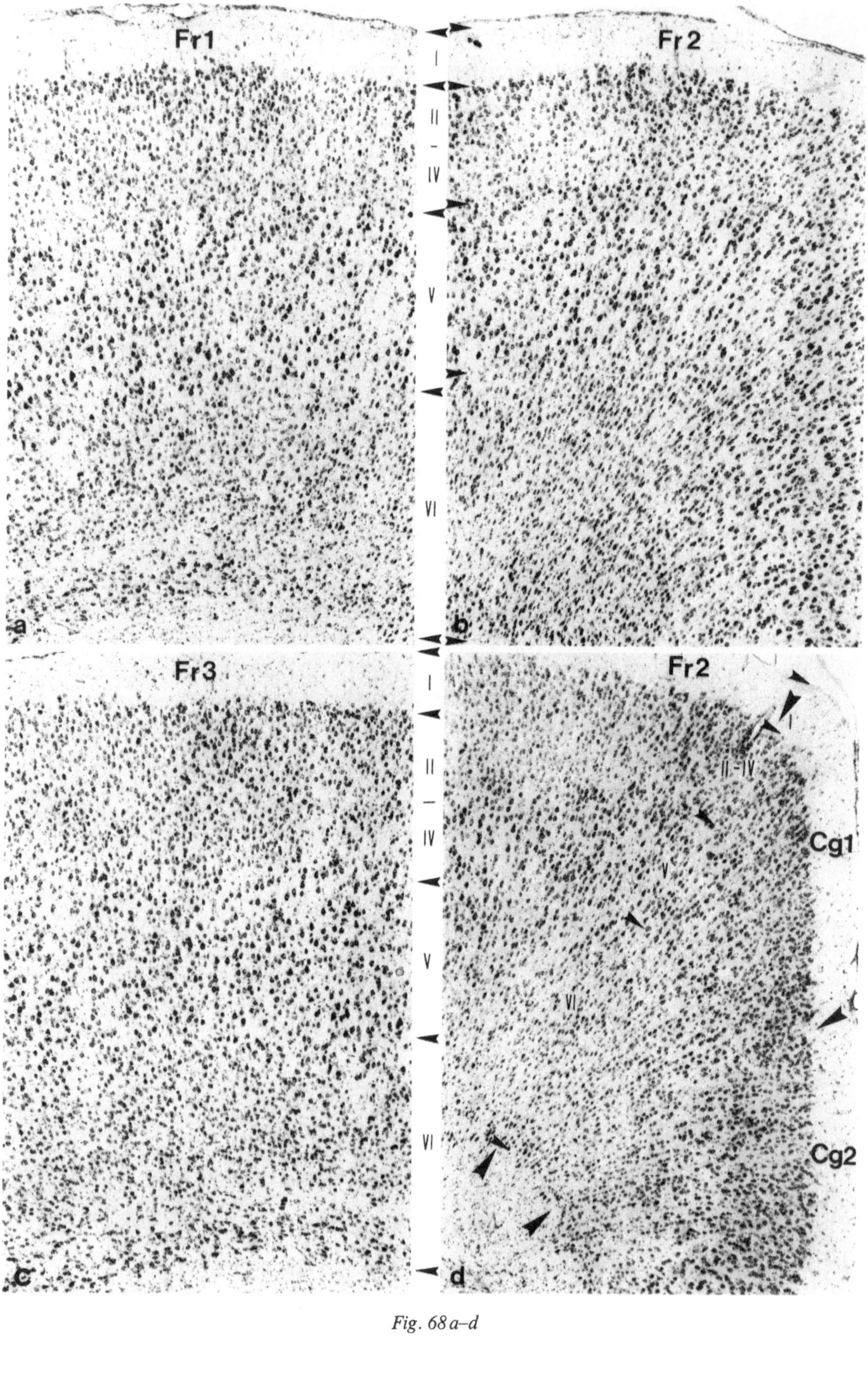

Fig. 68 a–d

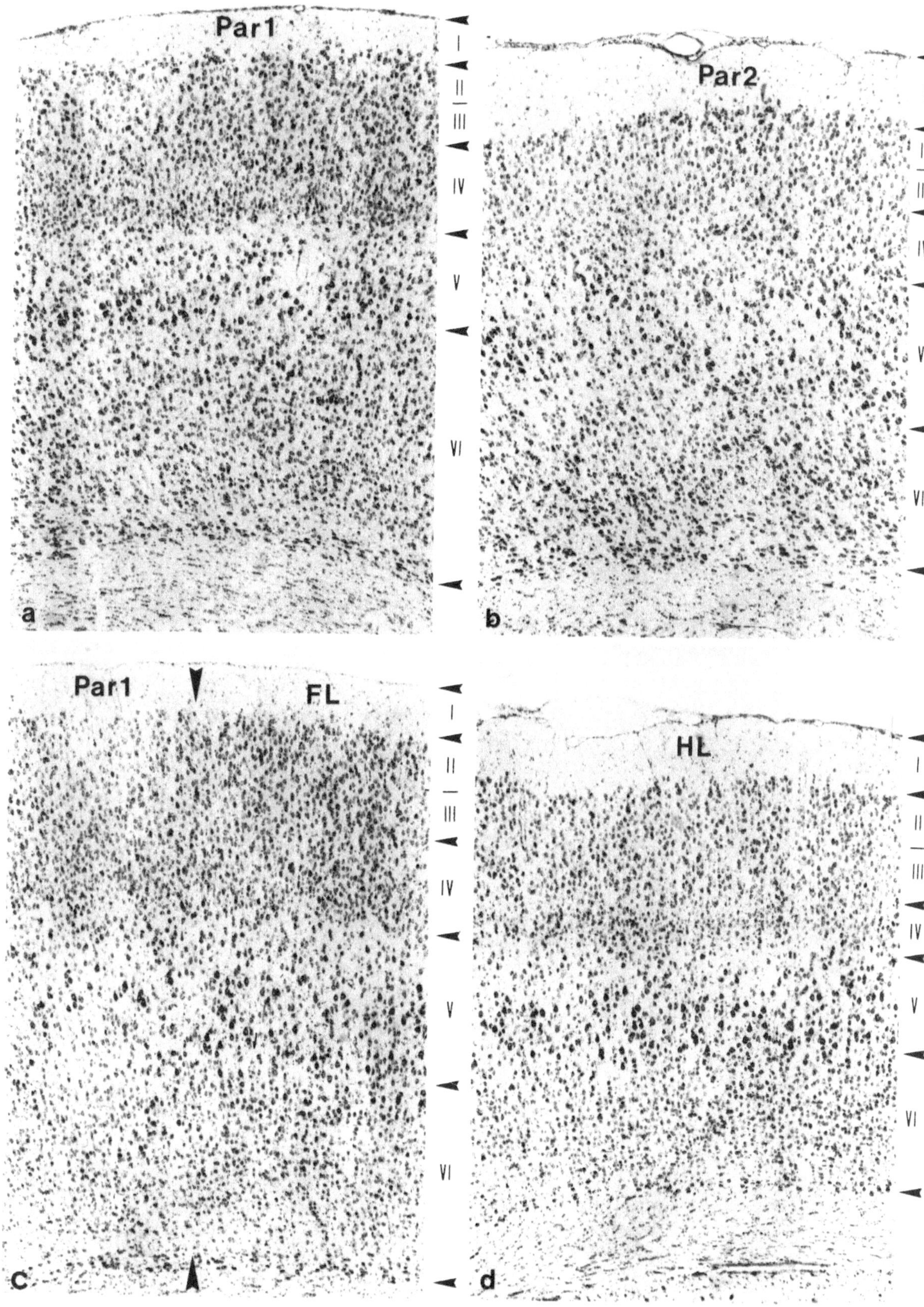

Par1
I
II
III
IV
V
VI
a
Par2
I
II
III
IV
V
VI
b
Par1
FL
I
II
III
IV
V
VI
c
HL
I
II
III
IV
V
VI
d

[99]

Fig. 69 a–d

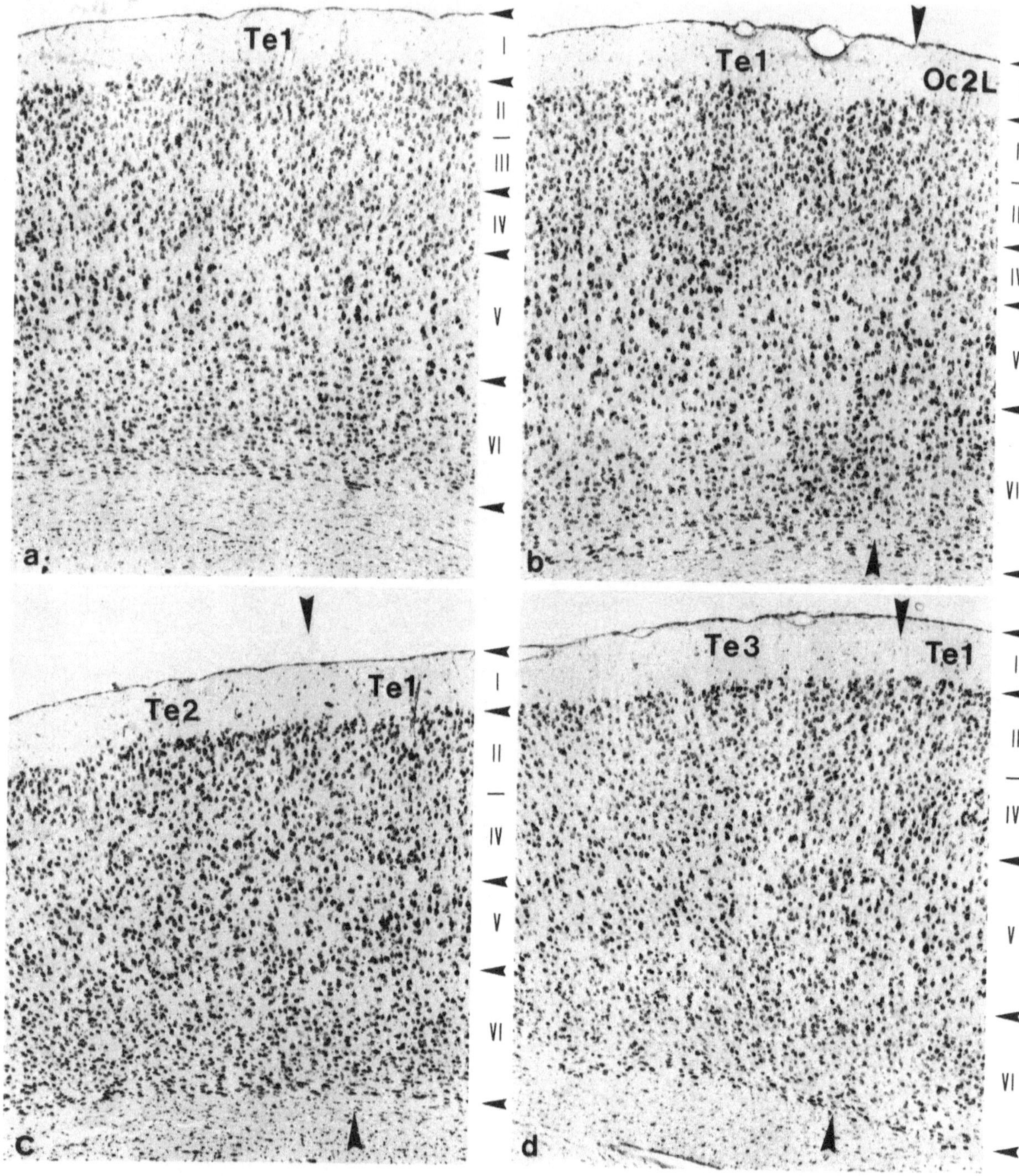

Fig. 70 a–d. Laminar pattern of cortical areas in the temporal cortex. Area Te1 (*a, b*) is shown at bregma −6.3 (*a*) and −5.3 (*b*) to demonstrate the structural inhomogeneity of this area (cf. Figs. 25 and 23). Area Te2 (*c*) is shown at bregma −6.3 (cf. Fig. 25) and area Te3 (*d*) at bregma −5.3 (cf. Fig. 23). Nissl; 64×

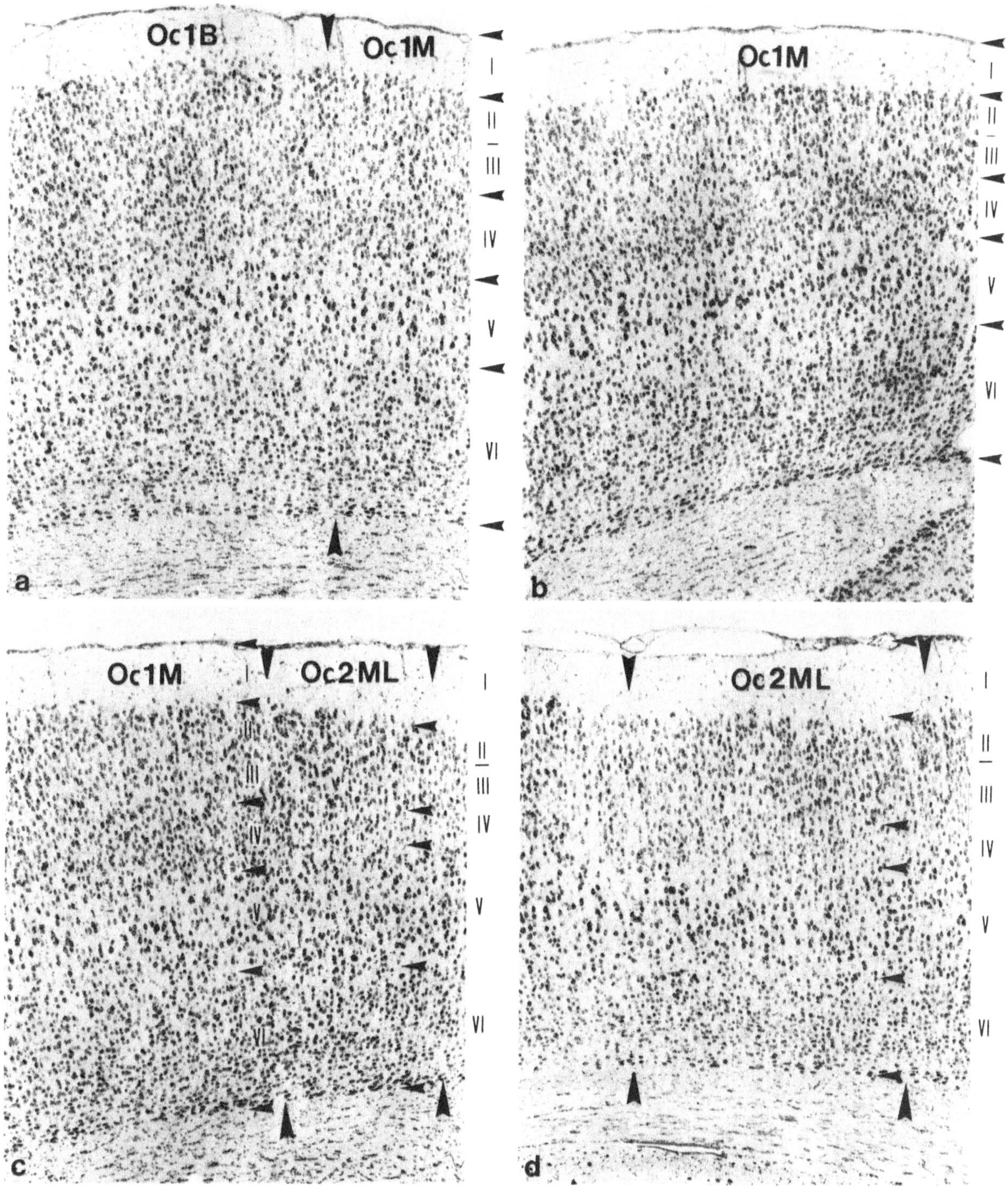

[101]

Fig. 71 a–d. Laminar pattern of cortical areas in the occipital cortex. Areas Oc1B (*a*) Oc1M (*a, b, c*) and Oc2ML (*c*) are shown at bregma −6.3 (cf. Fig. 25) area Oc1M (*b*) at bregma −7.3 (cf. Fig. 27), and area Oc2ML (*d*) at bregma −5.3 (cf. Fig. 23). Nissl; 64 ×

[102]

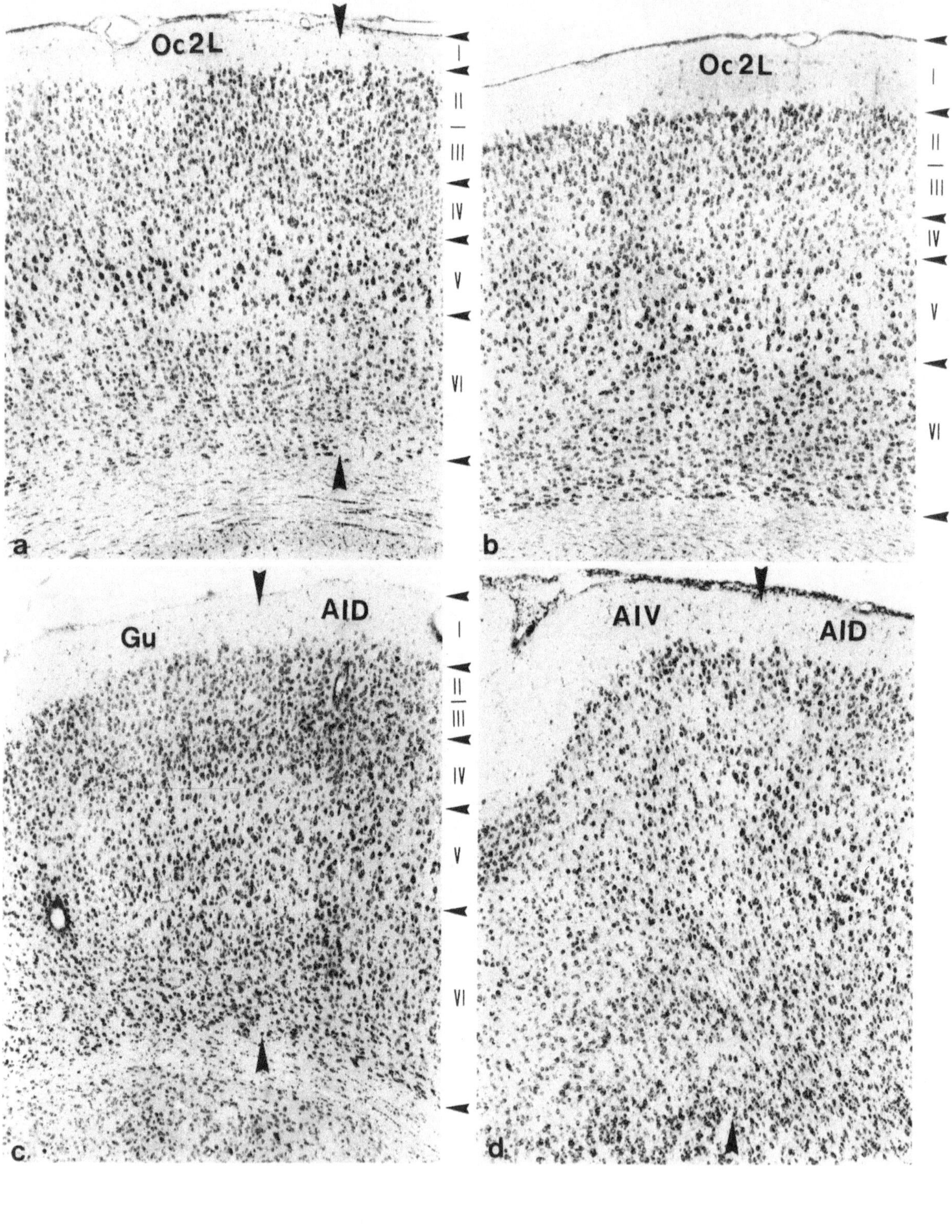

Fig. 72a–d. Laminar pattern of cortical areas in the occipital cortex. Area Oc2L (*a, b*) is shown at bregma −5.3 (*a*) and −7.3 (*b*) to demonstrate the structural inhomogeneity of this area (cf. Figs. 23 and 27). Areas Gu and AID (*c*) are shown at bregma −0.3 (cf. Fig. 13) and areas AID and AIV (*d*) at bregma 1.7 (cf. Fig. 9). Nissl; 64 ×

Aspects of Cortical Variability

This atlas is not intended to be a detailed study of interindividual variations, but efforts have been made to present some important aspects of this problem arising from the use of different rat strains. Figures 73–76 present the results of cortical mapping of different individuals from the same strain and of individuals from different strains (Han: Wistar strain, Lewis strain, hooded rats BDE/Han).

Figure 73 shows two cortical maps of the lateral aspect of the hemispheres superimposed. One cortical map is that of the rat presented in Figs. 2–43 and the other is that of two other male Wistar rats almost identical with the earlier rat in age and body weight. The basic topology of the cortical areas is identical in both brains. Moreover, the position and extension of each of these cortical areas is so well comparable that the areas in which the same field is found in both brains (white areas) are very much larger than the areas in which no identical fields are found (black areas). Upon superimposition of more brains from animals of the same strain the black areas presumably would increase in size though the preponderance of overlapping areas over nonoverlapping areas would persist. This emphasizes the interindividual variability within *one* rat strain, which is wide enough to lead us to doubt the significance and reliability of experiments based on uncritical confidence in stereotaxic coordinates mostly derived from only one brain, especially if the experiments involve regions near the borderline of a cortical area. On the other hand, Fig. 73 demonstrates that stereotaxic coordinates give important and reliable information about the position of a large proportion of a cortical area, which is not influenced by interindividual variation. Personal experiments with stereotaxic injections of HRP into the ares Oc1M, Oc1B, Oc2ML, and the border-region between Oc1M and Oc2ML have shown successful placement in all cases. These were checked by subsequent histological identification of the sites of these injections. In 2 of 11 cases the Oc1M injections were nearer the Oc1M/Oc2ML border than had been expected from the stereotaxic coordinates derived from Fig. 47. Nevertheless, the injections were confined to Oc1M.

Figures 74–76 present cortical maps resulting from the superimposition of the cortical maps of a Lewis rat, a hooded rat, and a Wistar rat brain. The maps show lateral, dorsal, and medial views of a model brain constructed from mean values for all these three brains. The outer contours of these brains were superimposed to allow graphic generation of the outer contour of the "average" brain. Analogous treatment of the areal borders of several cortical areas in these three brains yielded the delineation of core regions as stereotaxically definable minimal common areas in all three brains from these different strains. This type of map could prove valuable when a relatively reliable stereotaxic position of a cortical area is needed, e.g., at the beginning of tracer studies.

Acknowledgements

I would like to thank Dr. George Paxinos (School of Psychology, University of New South Wales), for allowing me to use his original histological sections
[104] for delineation of the cortical areas in Figs. 2–45 and for his many helpful and critical suggestions. I would also like to thank my coworkers at the Anatomical Institute, University of Cologne, Dr. Axel Schleicher, Dr. Andreas Wree, Peter Schwientek, and Frank-Helmut Rauch, for helpful comments, Ursula Blohm for preparing the histological slides, Inge Koch for the photographic work and Irene Schreiber for the drawings, Johannes Bernbeck for his help with the English version, and Maria Hoppe for typing the manuscript.

The investigations on which this book is based were supported by the Deutsche Forschungsgemeinschaft.

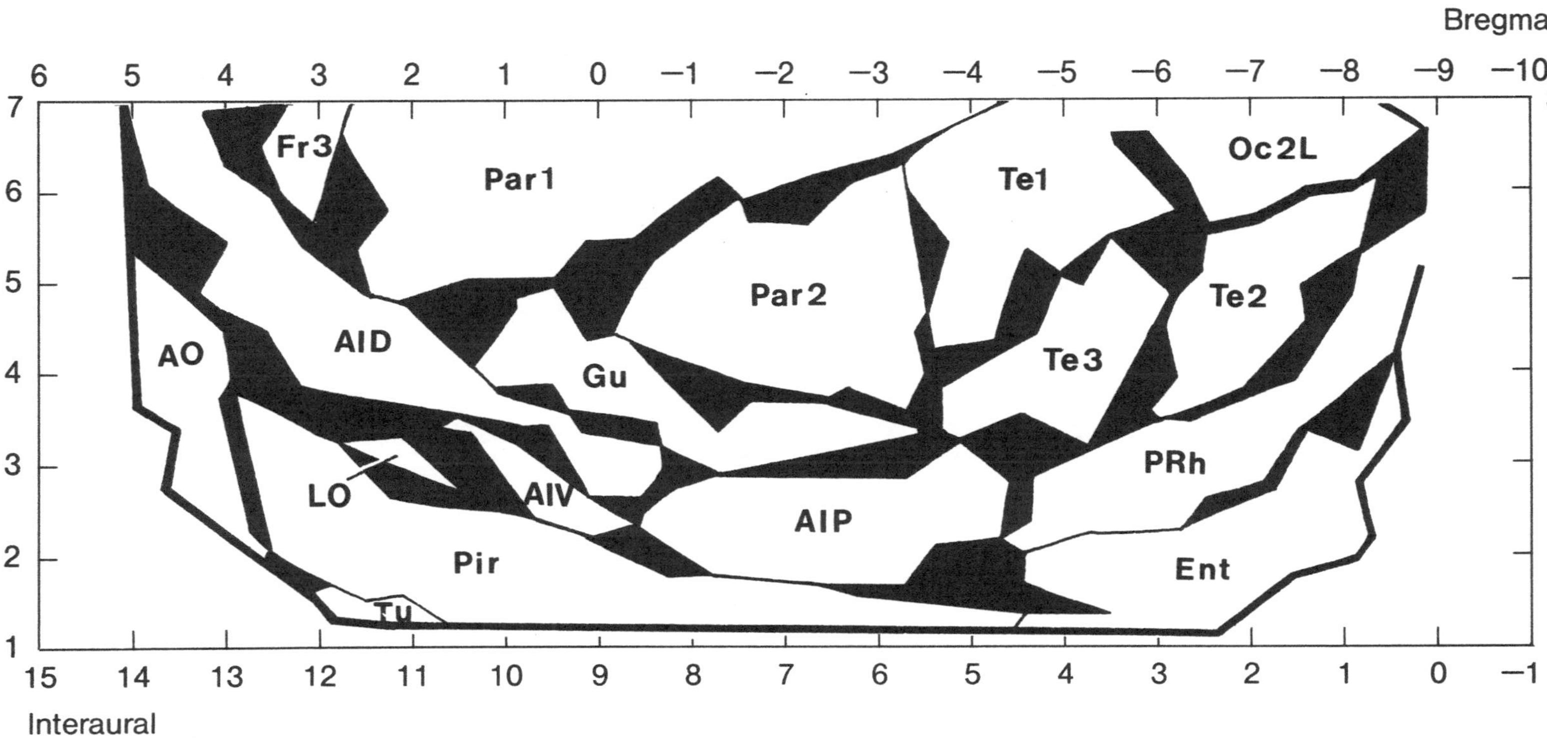
Bregma
6 5 4 3 2 1 0 −1 −2 −3 −4 −5 −6 −7 −8 −9 −10
7 6 5 4 3 2 1
3 4 5 6 7 8 9
15 14 13 12 11 10 9 8 7 6 5 4 3 2 1 0 −1
Interaural
Fr3
Par1
Te1
Oc2L
Par2
Te2
AO
AID
Gu
Te3
PRh
LO
AIV
AIP
Pir
Ent
Tu

Fig. 73

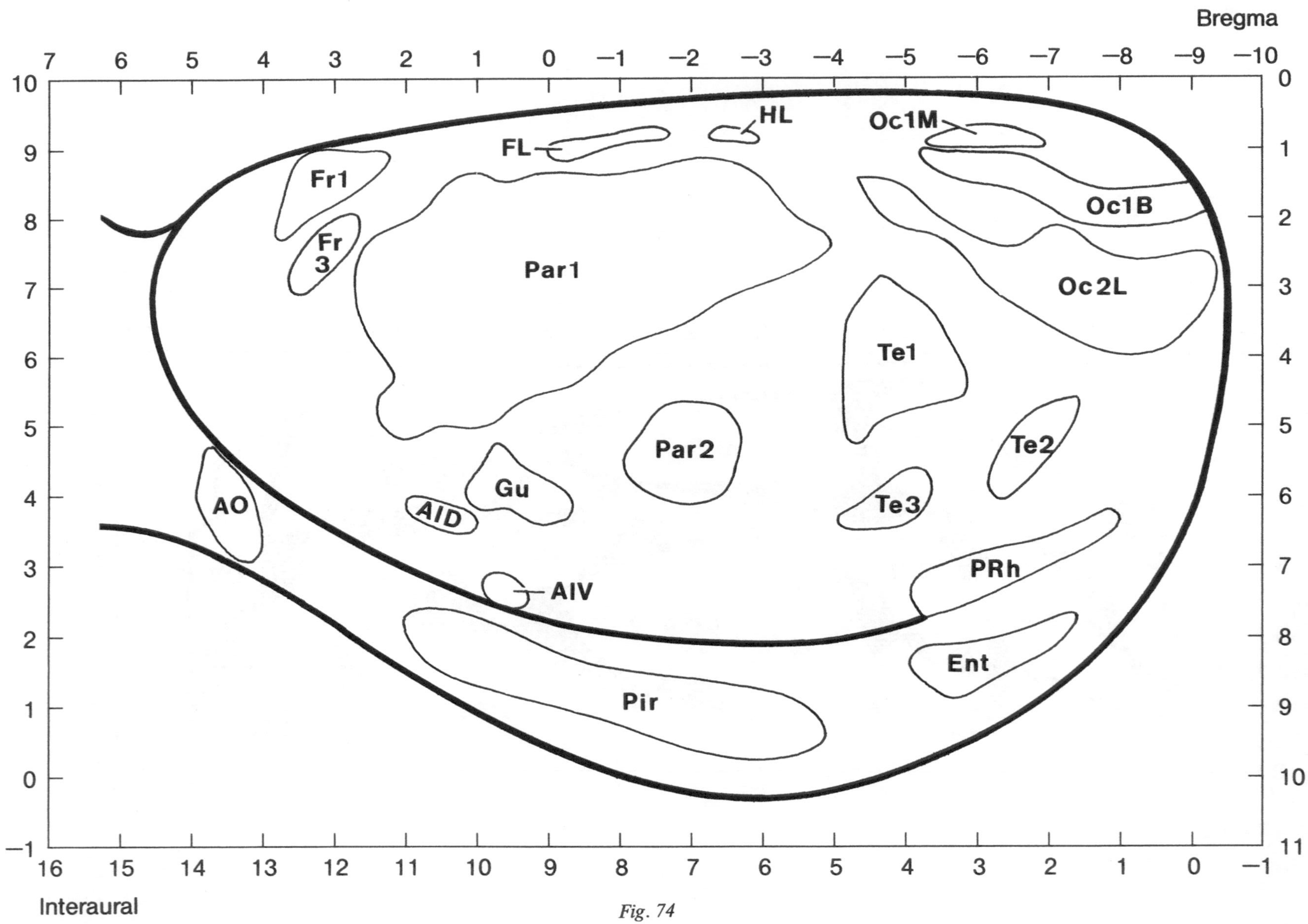
Bregma
Interaural
Oc1M
HL
FL
Fr1
Fr 3
Par1
Oc1B
Oc2L
Te1
Te2
Te3
Par2
Gu
AID
AO
AIV
PRh
Ent
Pir

Fig. 74

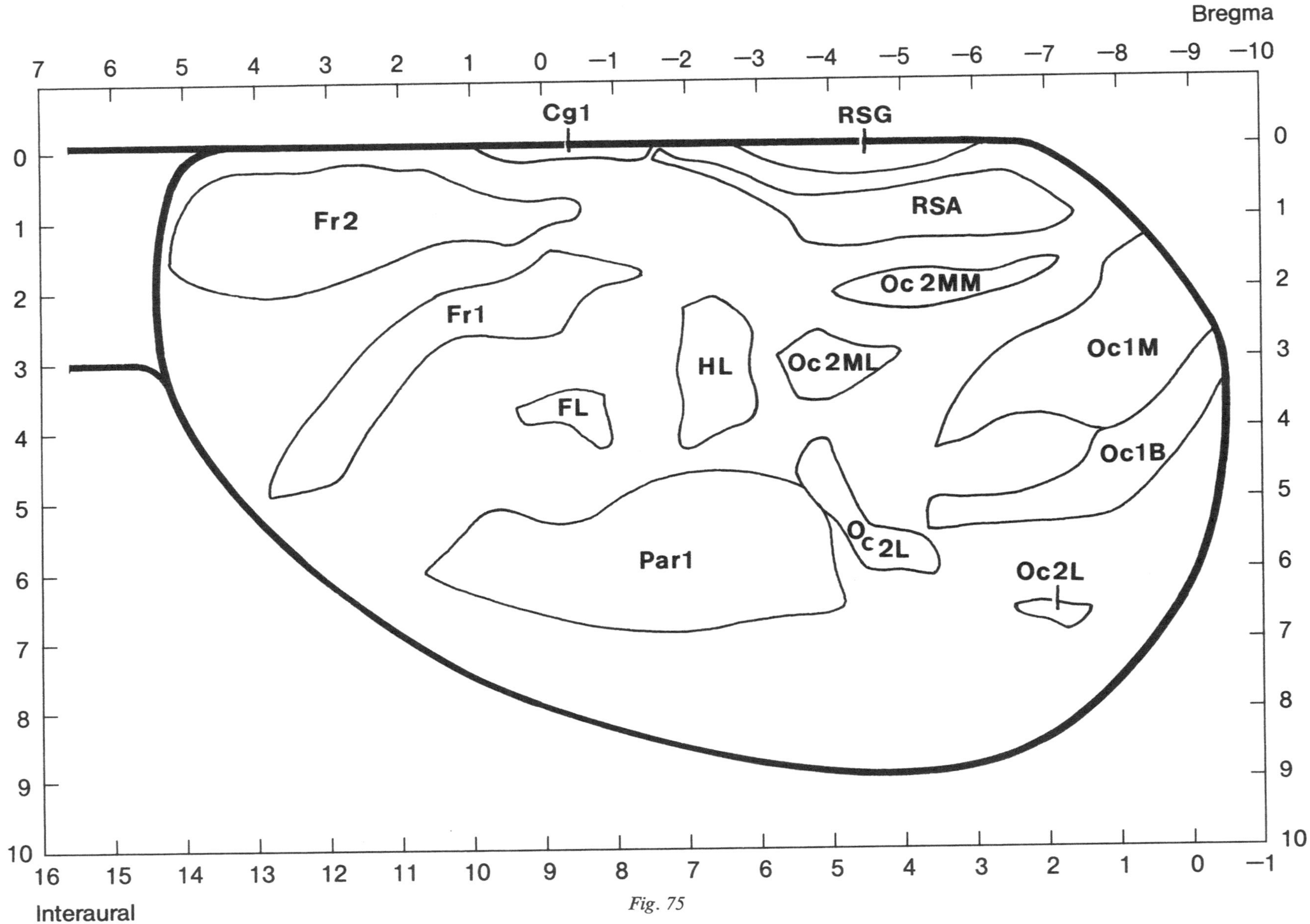
Bregma
Cg1
RSG
RSA
Fr2
Fr1
Oc 2MM
Oc1M
HL
Oc2ML
FL
Oc1B
Par1
Oc2L
Oc2L
Interaural

Fig. 75

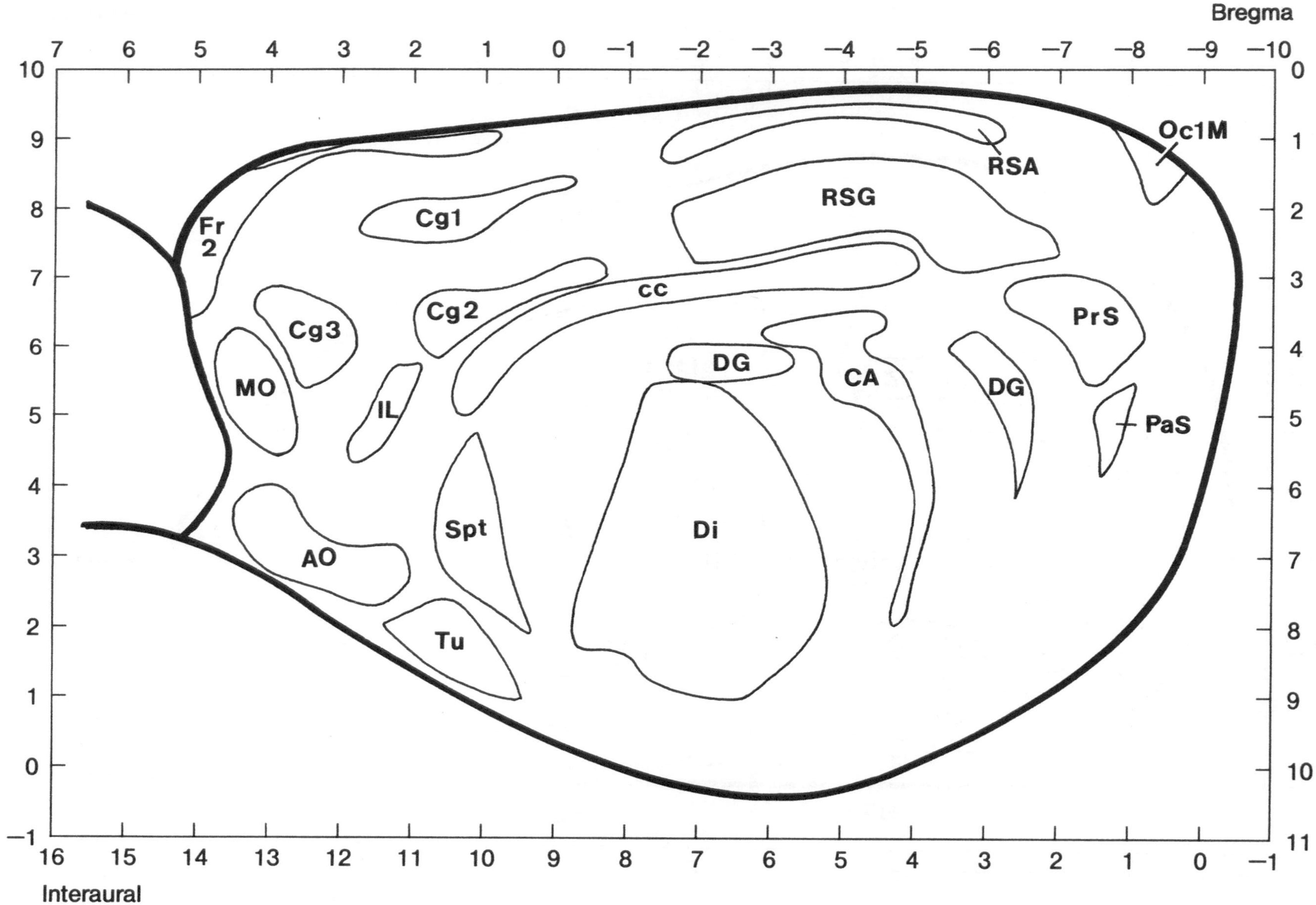
Bregma
Interaural
Oc1M
RSA
RSG
Cg1
Fr
2
cc
Cg2
Cg3
PrS
DG
CA
DG
MO
IL
PaS
Spt
Di
AO
Tu

Fig. 76

Index of Structures

The numbers are the numbers of the figures

References and Further Reading

Adams AD, Forrester JM (1968) The projection of the rat's visual field on the cerebral cortex. QJ Exp Physiol 53:327–336

Akers RM, Killackey HP (1978) Organization of cortico-cortical connections in the parietal cortex of the rat. J Comp Neurol 181:513–538

Astic L, Cattarelli M (1982) Metabolic mapping of functional activity in the rat olfactory system after a bilateral transection of the lateral olfactory tract. Brain Res 245:17–25

Bates CA, Killackey HP (1984) The emergence of a discretely distributed pattern of corticospinal projection neurons. 13:265–273

Beckstead RM (1976) Convergent thalamic and mesencephalic projections to the anterior medial cortex in the rat. J Comp Neurol 166:403–416

Beckstead RM (1979) An autoradiographic examination of cortico-cortical and subcortical projections of the mediodorsal-projection (prefrontal) cortex of the rat. J Comp Neurol 184:43–62

Benjamin RM, Akert K (1959) Cortical and thalamic areas involved in taste discrimination in the albino rat. J Comp Neurol 111:231–259

Benjamin RM, Pfaffmann C (1955) Cortical localization of taste in albino rat. J Neurophysiol 28:56–64

Benzinger H, Massopust LC (1983) Brain stem projections from cortical area 18 in the albino rat. Exp Brain Res 50:1–8

Berger B, Thierry AM, Tassin JP, Moyne MA (1976) Dopaminergic innervation of the rat prefrontal cortex: A fluorescence histochemical study. Brain Res 106:133–145

Bigl V, Woolf NJ, Butcher LL (1982) Cholinergic projections from the basal forebrain to frontal, parietal, temporal, occipital, and cingulate cortices: A combined fluorescent tracer and acetylcholinesterase analysis. Brain Res Bull 8:727–749

Braun JJ, Lasiter PS, Kiefer SW (1982) The gustatory neocortex of the rat. Physiol Psychol 10:13–45

Brodmann K (1909) Vergleichende Lokalisationslehre der Großhirnrinde in ihren Prinzipien dargestellt auf Grund des Zellenbaus. Barth, Leipzig

Brown LT (1974) Corticorubral projections in the rat. J Comp Neurol 154:149–168

Brown PA, Carman JB (1979) A specific pattern of connections from the sensorimotor region of the cerebral cortex to the thalamus in the rat. Acta Anat (Basel) 104:99–103

Catsman-Berrevoets CE, Kuypers HGJM (1981) A search for corticospinal collaterals of thalamus and mesencephalon by means of multiple retrograde fluorescent tracers in cat and rat. Brain Res 218:15–33

Caviness VS (1975) Architectonic map of neocortex of the normal mouse. J Comp Neurol 164:247–264

Cipolloni PB, Peters A (1979) The bilaminar and banded distribution of the callosal terminals in the posterior neocortex of the rat. Brain Res 176:33–47

Coleman J, Clerici WJ (1980) Extrastriate projections from thalamus to posterior occipital-temporal cortex in the rat. Brain Res 194:205–209

Cusick CG, Lund RD (1981) The distribution of callosal projection to the occipital visual cortex in rats and mice. Brain Res 214:239–259

Cusick CG, Lund RD (1982) Modification of visual callosal projections in rats. J Comp Neurol 212:385–398

Danner H, Pfister C (1981a) Untersuchungen zur Zytoarchitektonik des Tuberculum olfactorium der Ratte. J Hirnforsch 22:685–696

Danner H, Pfister C (1981b) Untersuchungen zur Zytoarchitektonik des Nucleus accumbens septi der Ratte. Anat Anz 150:264–280

Deacon TW, Eichenbaum H, Rosenberg P, Eckmann KW (1983) Afferent connections of the perirhinal cortex in the rat. J Comp Neurol 220:168–190

De Bruin JPC, van Oyen HGM, van de Poll N (1983) Behavioural changes following lesions of the orbital prefrontal cortex in male rats. Behav Brain Res 10:209–232

Divac I (1972) Neostriatum and functions of prefrontal cortex. Acta Neurobiol Exp (Warsz) 33:461–477

Divac I, Diemer NH (1980) Prefrontal system in the rat visualized by means of labeled deoxyglucose – further evidence for functional heterogeneity of the neostriatum. J Comp Neurol 190:1–13

Divac I, Björklund A, Lindvall O, Passingham RE (1978a) Converging projections from the mediodorsal thalamic nucleus and mesencephalic dopaminergic neurons to the neocortex in three species. J Comp Neurol 180:59–72

Divac I, Kosmal A, Björklund A, Lindvall O (1978b) Subcortical projections to the prefrontal cortex in the rat as revealed by the horseradish peroxidase technique. Neuroscience 3:785–796

Domesick VB (1969) Projections from the cingulate cortex in the rat. Brain Res 12:296–320

[114] Domesick VB (1972) Thalamic relationships of the medial cortex in the rat. Brain Behav Evol 6:457–483

Donaldson L, Hand PS, Morrison AR (1975) Cortical-thalamic relationships in the rat. Exp Neurol 47:448–458

Donoghue JP, Parham C (1983) Afferent connections of the lateral agranular field of the rat motor cortex. J Comp Neurol 217:390–404

Donoghue JP, Wise SP (1982) The motor cortex of the rat: Cytoarchitecture and microstimulation mapping. J Comp Neurol 212:76–88

Donoghue JP, Kerman KL, Ebner FF (1979) Evidence for two organizational plans within the somatic sensory-motor cortex of the rat. J Comp Neurol 183:647–664

Dow-Edwards D, Dam M, Peterson JM, Rapoport SI, London ED (1981) Effect of oxotremorine on local cerebral glucose utilization in motor system regions of the rat brain. Brain Res 226:281–289

Droogleever Fortuyn AB (1914) Cortical cell-lamination of the hemispheres of some rodents. Arch Neurol Psychol 6:221–354

Druga R (1982) Claustro-neocortical connections in the cat and rat demonstrated by HRP tracing technique. J Hirnforsch 23:191–202

Druga R, Syka J (1984) Ascending and descending projections to the inferior colliculus in the rat. Physiol Bohemoslov 33:31–42

Economo K von, Koskinas GN (1925) Die Cytoarchitektonik der Hirnrinde des erwachsenen Menschen. Springer, Wien Berlin

Eichenbaum H, Clegg RA, Feeley A (1983) Reexamination of functional subdivisions of the rodent prefrontal cortex. Exp Neurol 79:434–451

Espinoza SG, Thomas HC (1983) Retinotopic organization of striate and extrastriate visual cortex in the hooded rat. Brain Res 272:137–144

Fallon JH (1980) The islands of Calleja complex of rat basal forebrain. II: Connections of medium and large sized cells. Brain Res Bull 10:775–793

Fallon JH, Loughlin SE, Ribak CE (1983) The islands of Calleja complex of rat basal forebrain. III: Histochemical evidence for a striatopallidal system. J Comp Neurol 218:91–120

Fleischhauer K, Zilles K, Schleicher A (1980) A revised cytoarchitectonic

map of the neocortex of the rabbit (Oryctolagus cuniculus). Anat Embryol (Berl) 161:121–143
Friedman B, Price JL (1984) Fiber systems in the olfactory bulb and cortex: A study in adult and developing rats, using the Timm method with the light and electron microscope. J Comp Neurol 223:88–109
Gallyas F (1971) A principle for silver staining of tissue elements by physical development. Acta Morph Acad Sci Hung 19:57–71
Gallyas F (1979) Silver staining of myelin by means of physical development. Neurol Res 1:203–209
Gerfen CR, Clavier RM (1979) Neural inputs to the prefrontal agranular insular cortex in the rat: Horseradish peroxidase study. Brain Res Bull 4:347–353
Guldin WO, Markowitsch HJ (1983) Cortical and thalamic afferent connections of the insular and adjacent cortex of the rat. J Comp Neurol 215:135–153
Haberly LB, Price JL (1978) Association and commissural fiber systems of the olfactory cortex of the rat. I: Systems originating in the piriform cortex and adjacent areas. J Comp Neurol 178:711–740
Hall RD, Lindholm EP (1974) Organization of motor and somatosensory neocortex in the albino rat. Brain Res 66:23–38
Heller A, Hutchens JO, Kirby ML, Karapas F, Fernandez C (1979) Stereotaxic electrode placement in the neonatal rat. J Neurosci Methods 1:41–76
Herkenham M (1978) The connections of the nucleus reuniens thalami: Evidence for a direct thalamo-hippocampal pathway in the rat. J Comp Neurol 177:580–610
Herkenham M (1979) The afferent and efferent connections of the ventromedial thalamic nucleus in the rat. J Comp Neurol 183:487–518
Herkenham M (1980) Laminar organization of thalamic projections to the rat neocortex. Science 207:532–535
Hosoya Y, Matsushita M (1980) Cells of origin of the descending afferents to the lateral hypothalamic area in the rat, studied with the horseradish peroxidase method. Neurosci Lett 18:231–236
Hughes HC (1977) Anatomical and neurobehavioral investigations concerning the thalamo-cortical organization of the rat's visual system. J Comp Neurol 175:311–336
Isserhoff A, Schwartz ML, Decker JJ, Goldman-Rakic PS (1984) Columnar organization of callosal and associational projections from rat frontal cortex. Brain Res 293:213–223
Itaya SK, van Hoesen GW, Jenq C-B (1981) Direct retinal input to the limbic system of the rat. Brain Res 226:33–42
Ivy GO, Akers RM, Killackey HP (1979) Differential distribution of callosal projection neurons in the neonatal and adult rat. Brain Res 173:532–537
Jacobson S (1965) Intralaminar, interlaminar, callosal, and thalamocortical connections in frontal and parietal areas of the albino rat cerebral cortex. J Comp Neurol 124:131–146
Jacobson S (1970) Distribution of commissural axon terminals in the rat neocortex. Exp Neurol 28:193–205
Jacobson S, Trojanowski JQ (1975) Corticothalamic neurons and thalamocortical terminal fields. An investigation in rat using horseradish peroxidase and autoradiography. Brain Res 85:385–401
Jones EG, Leavitt RY (1974) Retrograde axonal transport and the demonstration of non-specific projections to the cerebral cortex and striatum from thalamic intralaminar nuclei in the rat, cat and monkey. J Comp Neurol 154:349–378
Jones EG, Porter R (1980) What is area 3a? Brain Res Rev 2:1–43
Kassel S (1982) Somatotopic organization of SI corticotectal projections in rats. Brain Res 231:247–255
Kelley AE, Domesick VB (1982) The distribution of the projection from the hippocampal formation to the nucleus accumbens in the rat: An anterograde and retrograde horseradish peroxidase study. Neuroscience

7:2321–2335
Killackey HP (1973) Anatomical evidence for cortical subdivisions based on vertically discrete thalamic projections from the ventral posterior nucleus to cortical barrels in the rat. Brain Res 51:326–331
Killackey HP, Leshin S (1975) The organization of specific thalamocortical projections to the posteromedial barrel subfield of the rat somatic sensory cortex. Brain Res 86:469–472
Killackey HP, Ryugo DK (1975) The organization of unspecific thalamic projections to the telencephalon of the rat. Anat Rec 181:393
Koelle GG, Friedenwald JS (1949) A histochemical method for localizing cholinesterase activity. Proc Soc Exp Biol Med 70:617–622
König JFR, Klippel RA (1963) The rat brain: A stereotaxic atlas of the forebrain and lower parts of the brain stem. Williams and Wilkins, Baltimore
Kosel KC, van Hoesen GW, Rosene DL (1982) Nonhippocampal cortical projections from the entorhinal cortex in the rat and rhesus monkey. Brain Res 244:201–213
Kosel KC, van Hoesen GW, Rosene DL (1983) A direct projection from the perirhinal cortex (area 35) to the subiculum in the rat. Brain Res 269:347–351
Krettek JE, Price JL (1977a) The cortical projections of the mediodorsal nucleus and adjacent thalamic nuclei in the rat. J Comp Neurol 171:157–192
Krettek JE, Price JL (1977b) Projections from the amygdaloid complex to the cerebral cortex and thalamus in the rat and cat. J Comp Neurol 172:687–722
Krieg WJS (1946a) Connections of the cerebral cortex. I. The albino rat. A. Topography of the cortical areas. J Comp Neurol 84:221–275
Krieg WJS (1946b) Connections of the cerebral cortex. I. The albino rat. B. Structure of the cortical areas. J Comp Neurol 84:276–323

Krieg WJS (1946c) Accurate placement of minute lesions in the brain of the albino rat. Q Bull Northwest Univ Med School 20:199–208
Krieg WJS (1947) Connections of the cerebral cortex. I. The albino rat. C. Extrinsic connections. J Comp Neurol 86:267–394
Kristt DA (1979) Somatosensory cortex: Acetylcholinesterase staining of barrel neuropil in the rat. Neurosci Lett 12:177–182
Lashley KS (1934) The mechanism of vision. VIII: The projection of the retina upon the cerebral cortex of the rat. J Comp Neurol 60:57–79
Lashley KS (1941) Thalamocortical connections of the rat's brain. J Comp Neurol 75:67–121
Lasiter PS, Glanzman DL (1983) Axon collaterals of pontine taste area neurons project to the posterior ventromedial thalamic nucleus and to the gustatory neocortex. Brain Res 258:299–304
Lasiter PS, Glanzmann DL, Mensah PA (1982) Direct connectivity between pontine taste areas and gustatory neocortex in rat. Brain Res 234:111–121
Le Messurier DH (1948) Auditory and visual areas of the cerebral cortex of the rat. Fed Proc 7:70–71
Leonard CM (1969) The prefrontal cortex of the rat. I. Cortical projection of the mediodorsal nucleus. II. Efferent connections. Brain Res 12:321–343
Leonard CM (1972) The connections of the dorsomedial nucleus. Brain Behav Evol 6:524–541
Leong SK (1983) Localizing the corticospinal neurons in neonatal, developing and mature albino rat. Brain Res 265:1–9
Lewis ME, Pert A, Pert CB, Herkenham M (1983) Opiate receptor localization in rat cerebral cortex. J Comp Neurol 216:339–358
Lewis PR (1961) The effect of varying the conditions in the Koelle method. Bibl Anat 2:11–20
Lidov HGW, Grzanna R, Molliver ME (1980) The serotonin innervation of the cerebral cortex in the rat – an immunohistochemical analysis. Neuroscience 5:207–227
Lindvall O, Björklund A, Divac I (1978) Organization of catecholamine neu-

rons projecting to the frontal cortex in the rat. Brain Res 142:1–24
Loughlin SE, Fallon JH (1984) Substantia nigra and ventral tegmental area projections to cortex: Topography and collateralization. Neuroscience 11:425–435
Luskin MB, Price JL (1983a) The topographic organization of associational fibers of the olfactory system in the rat, including centrifugal fibers to the olfactory bulb. J Comp Neurol 216:264–291
Luskin MB, Price JL (1983b) The laminar distribution of intracortical fibers originating in the olfactory cortex of the rat. J Comp Neurol 216:292–302
Markowitsch HJ, Pritzel M (1977) Comparative analysis of prefrontal learning functions in rats, cats and monkeys. Psychol Bull 84:817–837
Markowitsch HJ, Guldin WO (1983) Heterotopic interhemispheric cortical connections in the rat. Brain Res Bull 10:805–810
McDonald AJ (1983) Cytoarchitecture of the nucleus of the lateral olfactory tract: A Golgi study in the rat. Brain Res Bull 10:497–503
Michalski A, Patuwald R, Schulz E (1979) Qualitative Analyse der Neuronenstruktur der Regio retrosplenialis granularis der Ratte. J Hirnforsch 20:181–190
Miller MW, Vogt BA (1984a) Heterotopic and homotopic callosal connections in rat visual cortex. Brain Res 297:75–89
Miller MW, Vogt BA (1984b) Direct connections of rat visual cortex with sensory, motor, and association cortices. J Comp Neurol 226:184–202
Montero VM (1973) Evoked responses in the rat's visual cortex to contralateral, ipsilateral and restricted photic stimulation. Brain Res 53:192–196
Montero VM (1981) Comparative studies on the visual cortex. In: Woolsey CN (ed) Cortical sensory organization. Humana, Clifton, pp 33–81
Montero VM, Bravo H, Fernàndez V (1973a) Striate-peristriate cortico-cortical connections in the albino and gray rat. Brain Res 53:202–207
Montero VM, Rojas A, Torrealba F (1973b) Retinotopic organization of striate and peristriate visual cortex in the albino rat. Brain Res 53:197–201
Morrison JH, Molliver ME, Grzanna R, Coyle JT (1979) Noradrenergic innervation patterns in three regions of medial cortex: An immunofluorescence characterization. Brain Res Bull 4:849–857
Morrison JH, Molliver ME, Grzanna R, Coyle JT (1981) The intra-cortical trajectory of the coeruleo-cortical projection in the rat: A tangentially organized cortical afferent. Neuroscience 6:139–158
Nauta WJH, Bucher VM (1954) Efferent connections of the striate cortex in the albino rat. J Comp Neurol 100:257–285
Neafsey EJ, Sievert C (1982) A second forelimb motor area exists in rat frontal cortex. Brain Res 232:151–156
Olavarria J (1979) A horseradish peroxidase study of the projection from the latero-posterior nucleus to three lateral peristriate areas in the rat. Brain Res 173:137–141
Olavarria J, Montero VM (1981) Reciprocal connections between the striate cortex and extrastriate cortical visual areas in the rat. Brain Res 217:358–363
Olavarria J, Montero VM (1984) Relation of callosal and striate-extrastriate cortical connections in the rat: Morphological definition of extrastriate visual areas. Exp Brain Res 54:240–252
Olavarria J, Torrealba F (1978) The effect of acute lesions of the striate cortex on the retinotopic organization of the lateral peristriate cortex in the rat. Brain Res 151:386–391
Olavarria J, van Sluyters RC (1982) The projection from striate and extrastriate cortical areas to the superior colliculus in the rat. Brain Res 242:332–336
Olavarria J, van Sluyters RC (1983) Widespread callosal connections in infragranular visual cortex of the rat. Brain Res 279:233–237
Olavarria J, Mignano LR, van Sluyters RC (1982) Pattern of extrastriate visual areas connecting reciprocally with striate cortex in the mouse. Exp Neurol 78:775–779

Orsini JC, Monmaur P, Delacour J, Calas A (1982) The neocortex of rats behaving in a driven rotating wheel shows a band characterized by preferential uptake of deoxyglucose. Neurosci Lett 29:287–291
Palkovits M, Zaborszky L, Brownstein MJ, Fekete MIK, Herman JP, Kanyicska B (1979) Distribution of norepinephrine and dopamine in cerebral cortical areas of the rat. Brain Res Bull 4:593–601
Parnavelas JG, Lieberman AR, Webster KE (1977) Organization of neurons in the visual cortex, area 17, of the rat. J Anat 124:305–322
Parnavelas JG, Burne RA, Lin C-S (1983) Distribution and morphology of functionally identified neurons in the visual cortex of the rat. Brain Res 261:21–29
Patterson HA (1977) An anterograde degeneration and retrograde axonal transport study of the cortical projections of rat geniculate body. Ph. D. Dissertation, Boston University, Boston
Paxinos G, Watson C (1982) The rat brain in stereotaxic coordinates. Academic, Sydney
Pellegrino LJ, Cushman AJ (1979) A stereotaxic atlas of the rat brain. Plenum, New York
Perry VH (1980) A tectocortical visual pathway in the rat. Neuroscience 5:915–927
Peters A, Feldman ML (1976) The projection of the lateral geniculate nucleus to area 17 of the rat cerebral cortex. I: General description. J Neurocytol 5:63–84
Pidoux B, Verley R (1979) Projections of the cortical somatic I barrel subfield from ipsilateral vibrissae in adult rodents. Electroencephalogr Clin Neurol 46:715–726
Popoff N, Popoff I (1929) Allocortex bei der Ratte (*Mus decumanus*). J Psychol Neurol 39:257–332
[118] Post S, Mai JK (1980) Contribution to the amygdaloid projection field in the rat. A quantitative autoradiographic study. J Hirnforsch 21:199–225
Powell TPS, Cowan WM (1954) The connections of the midline and intralaminar nuclei of the thalamus of the rat. J Anat 88:307–319
Price JL (1977) Structural organization of the olfactory pathways, In: Le Magnen J, MacLeod P (eds) Olfaction and taste, vol VI. Information Retrieval, London, pp 87–95
Price TR, Webster KE (1972) The cortico-thalamic projection from the primary somatosensory cortex of the rat. Brain Res 44:636–640
Reep RL, Winans SS (1982) Efferent connection of dorsal and ventral agranular insular cortex in the hamster, Mesocricetus auratus. Neuroscience 7:2609–2635
Reep RL, Corwin JV, Hashimoto A, Watson RT (1984) Afferent connections of medial precentral cortex in the rat. Neurosci Lett 44:247–252
Ribak CE (1977) A note on the laminar organization of rat visual cortical projections. Exp Brain Res 27:413–418
Ribak CE, Fallon JH (1982) The island of Calleja complex of rat basal forebrain. I. Light and electron microscopic observations. J Comp Neurol 205:207–218
Ribak CE, Peters A (1975) An autoradiographic study of the projections from the lateral geniculate body of the rat. Brain Res 92:341–368
Robertson RT, Kaitz SS, Robards MJ (1980) A subcortical pathway links sensory and limbic system of the forebrain. Neurosci Lett 17:161–165
Rose JE, Woolsey CN (1948) The orbito-frontal cortex and its connections with the mediodorsal nucleus in rabbit, sheep and cat. Res Publ Assoc Res Nerv Ment Dis 27:210–232
Rose M (1912) Histologische Lokalisation der Großhirnrinde bei kleinen Säugetieren (Rodentia, Insectivora, Chiroptera). J Psychol Neurol 19:391–479
Rose M (1929) Cytoarchitektonischer Atlas der Großhirnrinde der Maus. J Psychol Neurol 40:1–51
Rose M (1931) Cytoarchitektonischer Atlas der Großhirnrinde des Kaninchens. J Psychol 43:353–440

Ryugo DK, Killackey HP (1974) Differential telencephalic projections of the medial and ventral divisions of the medial geniculate body of the rat. Brain Res 82:173–177
Ryzen M (1956) A microphotometric method of cell enumeration within the cerebral cortex of man. J Comp Neurol 104:233–245
Sanderson KJ, Welker W, Shames GM (1984) Reevaluation of motor cortex and of sensorimotor overlap in cerebral cortex of albino rats. Brain Res 292:251–260
Saper CB (1984) Organization of cerebral cortical afferent systems in the rat. II. Magnocellular basal nucleus. J Comp Neurol 222:313–342
Saporta S, Kruger L (1977) The organization of thalamocortical relay neurons in the transport of horseradish peroxidase. J Comp Neurol 147:187–208
Sarter M, Markowitsch HJ (1983) Convergence of basolateral amygdaloid and mediodorsal thalamic projections in different areas of the frontal cortex in the rat. Brain Res Bull 10:607–622
Sarter M, Markowitsch HJ (1984) Collateral innervation of the medial and lateral prefrontal cortex by amygdaloid, thalamic, and brain-stem neurons. J Comp Neurol 224:445–460
Schleicher A, Zilles K, Kretschmann H-J (1978) Automatische Registrierung und Auswertung eines Grauwertindex in histologischen Schnitten. Anat Anz [Suppl] 114:413–415
Schober W (1981) Efferente and afferente Verbindungen des Nucleus lateralis posterior thalami ("Pulvinar") der Albinoratte. Z Mikrosk Anat Forsch 95:827–844
Schober W, Gruschka H (1983) Zur subkortikalen Projektion des visuellen Kortex bei der adulten Albinoratte und während der postnatalen Entwicklung. Z Mikrosk Anat Forsch 97:797–815
Schober W, Winkelmann E (1977) Die geniculo-corticale Projektion bei Albinoratten. J Hirnforsch 18:1–20
Schober W, Lüth H-J, Gruschka H (1976) Die Herkunft afferenter Axone im striären Kortex der Albinoratte: Eine Studie mit Meerrettich-Peroxidase. Z Mikrosk Anat Forsch 90:399–415

Schulz E, Schwartz A, Schober W, Holz L (1983) Zur neuronalen Lokalisation thalamischer Afferenzen in der Regio cingularis der Ratte. J Hirnforsch 24:535–544
Schwientek P (1985) Quantitative Analyse des arealen und laminären Aufbaus des Isocortex der Ratte. Medical dissertation, University of Cologne
Schwob JE, Price JL (1978) The cortical projection of the olfactory bulb: Development in fetal and neonatal rats correlated with quantitative variations in adult rats. Brain Res 151:369–374
Sefton AJ, Mackay-Sim A, Baur LS, Cotte LJ (1981) Cortical projections to visual centres in the rat: An HRP study. Brain Res 215:1–13
Settlage PH, Bingham WG, Suckle HM, Borge AF, Woolsey CN (1949) The pattern of localization in the motor cortex of the rat. Fed Proc 8:144
Sharp FR, Kauer JS, Shepherd GM (1977) Laminar analysis of 2-deoxyglucose uptake in olfactory bulb and olfactory cortex of rabbit and rat. J Neurophysiol 40:800–813
Sherwood NM, Timiras PS (1970) A stereotaxic atlas of the developing rat brain. University of California Press, Berkeley
Shiosaka S, Tohyama M, Takagi H, Takahshy Y, Saitoh Y, Sakumoto T, Nakagawa H, Shimizu N (1980) Ascending and descending components of the medial forebrain bundle in the rat as demonstrated by the horseradish peroxidase blue reaction. Exp Brain Res 39:377–388
Sofroniew MV, Eckenstein F, Thoenen H, Cuello AC (1982) Topography of choline acetyltransferase-containing neurons in the forebrain of the rat. Neurosci Lett 33:7–12
Sokoloff L, Reivich M, Kennedy C, des Rosiers MH, Patlak CS, Sakurada KD, Shinohara M (1977) The ^{14}C deoxyglucose method for the measurement of local cerebral glucose utilization: theory, procedure, and normal values in the conscious and anesthetized albino rat. J Neurochem

28:897–916

Stephan H (1975) Allocortex. In: Bargmann W (ed) Nervensystem. Springer, Berlin Heidelberg, New York (Handbuch der mikroskopischen Anatomie des Menschen, vol 4/9)

Svetukhina VM (1962) Cytoarchitectonics of cerebral neocortex in rodents (Albino rat). Arch Anat Histol Embryol (Strasb) 42:31–45

Swanson LW (1981) A direct projection from Ammon's horn to prefrontal cortex in the rat. Brain Res 217:150–154

Swanson LW, Cowan WM (1977) An autoradiographic study of the organization of the efferent connections of the hippocampal formation in the rat. J Comp Neurol 172:749–784

Swanson LW, Cowan WM (1979) The connections of the septal region in the rat. J Comp Neurol 186:621–656

Swanson LW, Cowan WM, Jones EG (1974) An autoradiographic study of the efferent connections of the ventral lateral geniculate nucleus in the albino rat and cat. J Comp Neurol 156:143–164

Toga AW, Collins RC (1981) Metabolic response of optic centers to visual stimuli in the albino rat: Anatomical and physiological considerations. J Comp Neurol 199:443–464

Tsang Y-C (1937) Visual centers in blinded rats. J Comp Neurol 66:211–261

Ungerstedt U (1971) Stereotaxic mapping of the monoamine pathways in the rat brain. Acta Physiol Scand [Suppl] 367:1–48

Van der Kooy D, McGinty JF, Koda LY, Gerfen CR, Bloom FE (1982) Visceral cortex: A direct connection from prefrontal cortex to the solitary nucleus in rat. Neurosci Lett 33:123–127

Veazey RB, Severin CM (1982) Afferent projections to the deep mesencephalic nucleus in the rat. J Comp Neurol 204:134–150

Verrier M, Giachetti I, Leveteau J, MacLeod P (1977) Mapping of functional olfactory pathways by autoradiography with (^{14}C) 2-deoxyglucose. In: LeMagnen J, MacLeod P (eds) Olfaction and taste, vol VI. Information Retrieval, London, pp 208

Vogt BA, Miller MW (1983) Cortical connections between rat cingulate cortex and visual, motor and postsubicular cortices. J Comp Neurol 216:192–210

Vogt BA, Peters A (1981) Form and distribution of neurons in rat cingulate cortex: Area 32, 24, and 29. J Comp Neurol 195:603–625

Vogt BA, Rosene DL, Peters A (1981) Synaptic termination of thalamic and callosal afferents in cingulate cortex of the rat. J Comp Neurol 201:265–283

Volkmann U von (1926) Vergleichende Untersuchungen an der Rinde der "motorischen" und "Sehregion" von Nagetieren. Anat Anz [Suppl] 61:234–243

Webster KE (1961) Cortico-striate interrelations in the albino rat. J Anat 95:532–544

Welker C (1971) Microelectrode delineation of fine grain somatotopic organization of SmI cerebral neocortex in albino rat. Brain Res 26:259–275

Welker C (1976) Receptive fields of barrels in the somatosensory neocortex of the rat. J Comp Neurol 166:173–190

Welker C, Sinha M (1972) Somatotopic organization of SmII cerebral neocortex in albino rat. Brain Res 37:132–136

Welker W, Sanderson KJ, Shambes GM (1984) Patterns of afferent projection to transitional zones in the somatic sensorimotor cerebral cortex of albino rats. Brain Res 292:261–267

Werner L, Hedlich A, Winkelmann E, Brauer K (1979) Versuch einer Identifizierung von Nervenzellen des visuellen Kortex der Ratte nach Nissl- und Golgi-Kopsch-Darstellung. J Hirnforsch 20:121–139

Werner L, Wilke A, Blödner R, Winkelmann E, Brauer K (1982) Topographical distribution of neuronal types in the albino rat's area 17. A qualitative and quantitative Nissl study. Z Mikrosk Anat Forsch 96:433–453

Wiesendanger R, Wiesendanger M (1982a) The corticopontine system in the rat. I: Mapping of corticopontine neurons. J Comp Neurol 208:215–226

Wiesendanger R, Wiesendanger M (1982b) The corticopontine system in the rat. II: The projection pattern. J Comp Neurol 208:227–238
Winkelmann E, Kunz G, Winkelmann A (1972) Untersuchungen zur laminären Organisation des Cortex cerebri der Ratte unter besonderer Berücksichtigung der Sehrinde (Area 17). Z Mikrosk Anat Forsch 85:353–364
Winkelmann E, Hedlich A, Lüth H-J, Brauer K (1981) Zur neuronalen Organisation der Sehrinde. Z Mikrosk Anat Forsch 95:369–380
Wise SP (1975) The laminar organization of certain afferent and efferent fiber systems in the rat somatosensory cortex. Brain Res 90:139–142
Wise SP, Jones EG (1976) The organization and postnatal development of the commissural projection of the somatic sensory cortex of the rat. J Comp Neurol 168:313–343
Wise SP, Jones EG (1977) Cells of origin and terminal distribution of descending projections of the rat somatic sensory cortex. J Comp Neurol 175:129–158
Wise SP, Murray EA, Coulter JD (1979) Somatotopic organization of corticospinal and corticotrigeminal neurons in the rat. Neuroscience 4:65–78
Wolf G (1968) Projections of thalamic and cortical gustatory areas in the rat. J Comp Neurol 132:519–530
Woolf NJ, Eckenstein F, Butcher LL (1983) Cholinergic projections from the basal forebrain to the frontal cortex: a combined fluorescent tracer and immunohistochemical analysis in the rat. Neurosci Lett 40:93–98
Woolsey CN, LeMessurier DH (1948) The pattern of cutaneous representation in the rat's cerebral cortex. Fed Proc 7:137–138
Wree A, Zilles K, Schleicher A (1981) A quantitative approach to cytoarchitectonics. VII: The areal pattern of the cortex of the guinea pig. Anat Embryol (Berl) 162:81–103
Wree A, Schleicher A, Zilles K (1982) Estimation of volume fractions in nervous tissue with an image analyzer. J Neurosci Methods 6:29–43

Wree A, Zilles K, Schleicher A (1983) A quantitative approach to cytoarchitectonics. VIII: The areal pattern of the cortex of the albino mouse. Anat Embryol (Berl) 166:333–353
Wünscher W, Schober W, Werner L (1965) Architektonischer Atlas vom Hirnstamm der Ratte. Hirzel, Leipzig
Wyss JM (1981) An autoradiographic study of the efferent connections of the entorhinal cortex in the rat. J Comp Neurol 199:495–512
Wyss JM, Sripanidkulchai K (1983) The indusium griseum and anterior hippocampal continuation in the rat. J Comp Neurol 219:251–272
Wyss JM, Sripanidkulchai K (1984) The topography of the mesencephalic and pontine projections from the cingulate cortex of the rat. Brain Res 293:1–15
Yanai J (1979) Strain and sex differences in the rat brain. Acta Anat (Basel) 103:150–158
Záborszky L, Wolff JR (1982) Distribution patterns and individual variations of callosal connections in the albino rat. Anat Embryol (Berl) 165:213–232
Zilles K, Wree A (1985) Cortex: Areal and laminar structure. In: Paxinos G (ed) Forebrain and midbrain. Academic, Sydney (The rat nervous system, vol 1)
Zilles K, Zilles B, Schleicher A (1980) A quantitative approach to cytoarchitectonics. VI: The areal pattern of the cortex of the albino rat. Anat Embryol (Berl) 159:335–360
Zilles K, Stephan H, Schleicher A (1982) Quantitative cytoarchitectonics of the cerebral cortices of several prosimian species. In: Armstrong E, Falk D (eds) Primate brain evolution. Plenum, New York, pp 177–201
Zilles K, Wree A, Schleicher A, Divac I (1984) The monocular and binocular subfields of the rat's primary visual cortex. A quantitative morphological approach. J Comp Neurol 226:391–402
Zimmerman EA, Chambers WW, Liu CN (1964) An experimental study of the anatomical organization of the corticobulbar system in the albino rat. J Comp Neurol 123:301–324

LIST OF ABBREVIATIONS

Abbreviations used in the figures are listed in alphabetical order; each abbreviation is followed by the full name of the structure

AA	anterior amygdaloid area
ac	anterior commissure
aca	anterior commissure, anterior part
Acb	accumbens nucleus
aci	anterior commissure, intrabulbar part
ACo	anterior cortical amygdaloid nucleus
acp	anterior commissure, posterior part
AHi	amygdalohippocampal area
AID	agranular insular cortex, dorsal part (claustrocortex)
AIP	agranular insular cortex, posterior part (claustrocortex)
AIV	agranular insular cortex, ventral part (claustrocortex)
alv	alveus of the hippocampus
Amg	amygdaloid body
AO	anterior olfactory nucleus (retrobulbar region)
AOB	accessory olfactory bulb
B	cells of the basal nucleus of Meynert equivalent
BL	basolateral amygdaloid nucleus
BLV	basolateral amygdaloid nucleus, ventral part
BM	basomedial amygdaloid nucleus
BOl	olfactory bulb
CA	Ammon's horn
CA1	field CA1 of Ammon's horn
CA2	field CA2 of Ammon's horn
CA3	field CA3 of Ammon's horn
CA4	field CA4 of Ammon's horn (hilus fasciae dentatae)
Cb	cerebellum
cc	corpus callosum
Ce	central amygdaloid nucleus
CeL	central amygdaloid nucleus, lateral part
CeM	central amygdaloid nucleus, medial part
CG	central (periaqueductal) gray
Cg1	cingulate cortex, area 1 (medial prefrontal cortex)
Cg2	cingulate cortex, area 2 (medial prefrontal cortex)
Cg3	cingulate cortex, area 3 (medial prefrontal cortex)
CGD	central gray, dorsal part
Cl	claustrum
cp	cerebral peduncle
CPu	caudate putamen (striatum)
CxA	cortex-amygdala transition zone
df	dorsal fornix
DG	dentate gyrus
dhc	dorsal hippocampal commissure
Di	diencephalon
DPC	dorsal peduncular cortex
En	endopiriform nucleus
Ent	entorhinal area
Epi	epiphysis cerebri
f	fornix
fi	fimbria of the hippocampus
FL	forelimb area
fmi	forceps minor
fmj	forceps major
fr	fasciculus retroflexus
Fr1	frontal cortex, area 1 (primary motor cortex)
Fr2	frontal cortex, area 2
Fr3	frontal cortex, area 3
FStr	fundus striati
gcc	genu of the corpus callosum
GP	globus pallidus
Gu	gustatory cortex
HDB	nucleus of the horizontal limb of the diagonal band (Broca)
HL	hindlimb area
Hy	hypothalamus
I	intercalated nuclei of the amygdala
IC	inferior colliculus
ic	internal capsule
IG	indusium griseum
IL	infralimbic area of the medial frontal cortex
La	lateral amygdaloid nucleus
LO	lateral orbital area
lo	lateral olfactory tract
LOT	nucleus of the lateral olfactory tract
LOTD	nucleus of the lateral olfactory tract, dorsal part
LS	lateral septal nucleus
LSD	lateral septal nucleus, dorsal part
LSI	lateral septal nucleus, intermediate part
LSV	lateral septal nucleus, ventral part
mcp	middle cerebellar peduncle
Me	medial amygdaloid nucleus
MG	medial geniculate nucleus
ml	medial lemniscus
mlf	medial longitudinal fasciculus
MO	medial orbital area
MP	medial mammillary nucleus, posterior part
MS	medial septal nucleus
mt	mammillothalamic tract
Oc1B	occipital cortex, area 1 binocular part (primary visual cortex)
Oc1M	occipital cortex, area 1 monocular part (primary visual cortex)
Oc2L	occipital cortex, area 2 lateral part
Oc2ML	occipital cortex, area 2 mediolateral part
Oc2MM	occipital cortex, area 2 mediomedial part
opt	optic tract
ox	optic chiasm
Par1	parietal cortex, area 1 (primary somatosensory cortex)
Par2	parietal cortex, area 2 (supplementary somatosensory cortex)
PaS	parasubiculum
pc	posterior commissure
Pir	prepiriform cortex ("primary olfactory cortex")
PLCo	posterolateral cortical amygdaloid nucleus
PMCo	posteromedial cortical amygdaloid nucleus
Pn	pontine nuclei
PRh	perirhinal area
PrS	presubiculum
py	pyramidal tract
RSA	agranular retrosplenial cortex
RSG	granular retrosplenial cortex
S	subiculum
SC	superior colliculus
scc	splenium of the corpus callosum
scp	superior cerebellar peduncle
SFI	septofimbrial nucleus
SFO	subfornical organ
SHi	septohippocampal nucleus
SHy	septohypothalamic nucleus
sm	stria medullaris
sox	supraoptic decussation
Spt	septum
st	stria terminalis
str	superior thalamic radiation
Te1	temporal cortex, area 1 (primary auditory cortex)
Te2	temporal cortex, area 2
Te3	temporal cortex, area 3
tfp	transverse fibers of the pons
Th	thalamus
TS	triangular septal nucleus (interstitial nucleus of the ventral hippocampal commissure)
TT	taenia tecta
Tu	olfactory tubercle
VDB	nucleus of the vertical limb of the diagonal band (Broca)
vhc	ventral hippocampal commissure
VLO	ventrolateral orbital area
VO	ventral orbital area
xscp	decussation of the superior cerebellar peduncle